Contents

05/07

UNIVERSITY OF
WOLVERHAMPTON

Introduction to Engineering Ethics

Roland Schinzinger
Department of Electrical & Computer Engineering
University of California, Irvine

Mike W. Martin
Department of Philosophy
Chapman University

Boston   Burr Ridge, IL   Dubuque, IA   Madison, WI   New York   San Francisco   St. Louis
Bangkok   Bogotá   Caracas   Lisbon   London   Madrid
Mexico City   Milan   New Delhi   Seoul   Singapore   Sydney   Taipei   Toronto

For Shirley Barrows Price,
and in memory of Jane Harris Schinzinger.

Roland Schinzinger

For Shannon,
and in memory of my parents,
Ruth L. Martin and Theodore R. Martin.

Mike W. Martin

McGraw-Hill Higher Education

*A Division of The **McGraw-Hill** Companies*

INTRODUCTION TO ENGINEERING ETHICS
Copyright © 2000 by The McGraw-Hill Companies, Inc. All rights reserved. Printed in the United States of America. Except as permitted under the United States Copyright Act of 1976, no part of this publication may be reproduced or distributed in any form or by any means, or stored in a database retrieval system, without the prior written permission of the publisher.

This book is printed on acid-free paper.

4 5 6 7 8 9 0 FGR/FGR 0 9 8 7 6 5 4 3

ISBN 0-07-233959-4

Publisher: *Thomas Casson*
Executive editor: *Eric M. Munson*
Editorial coordinator: *Michael Jones*
Senior marketing manager: *John T. Wannemacher*
Project manager: *Paula Krauza*
Senior production supervisor: *Heather D. Burbridge*
Freelance design coordinator: *Pam Verros*
Cover designer: *JoAnne Schopler*
Cover photographs: © *Photodisc*
Compositor: *Lachina Publishing Services*
Typeface: *10/12 Century Schoolbook*
Printer: *Quebecor Printing Book Group/Fairfield*

Library of Congress Cataloging-in-Publication Data

Schinzinger, Roland.
    Introduction to engineering ethics / Roland Schinzinger, Mike W. Martin.
        p.   cm. — (McGraw-Hill's BEST—basic engineering series and tools)
    ISBN 0-07-233959-4 (softcover)
    1. Engineering ethics.   I. Martin, Mike W., 1946– .   II. Title.
III. Series.
TA157.S382   2000
174'.962—dc21   99-39643

http://www.mhhe.com

About the Authors

Roland Schinzinger and Mike W. Martin participated as an engineer–philosopher team in the National Project on Philosophy and Engineering Ethics from 1978 to 1980. Since then they have coauthored articles, team-taught courses, and together given presentations to audiences of engineers and philosophers. In 1992, they received the Award for Distinguished Literary Contributions Furthering Engineering Professionalism from The Institute of Electrical and Electronics Engineers, United States Activities Board.

Introduction to Engineering Ethics is a condensed and updated version of their book, *Ethics in Engineering,* which has been published in several editions and translations.

ROLAND SCHINZINGER, born and raised in Japan, worked in industry there for several years and upon emigrating to the U.S. in 1949 earned his B.S. (1953) and M.S. (1954) in electrical engineering at the University of California at Berkeley. He then worked at Westinghouse Electric (1954–58) on the design of large generators and also lectured at the University of Pittsburgh. After teaching engineering for five years in Istanbul (Robert College/Bosporus University), he obtained his Ph.D. at the University of California, Berkeley, then became a founding faculty member at the new University of California, Irvine campus. There he served in various capacities until retirement in 1993. Professor Schinzinger's research interests include energy systems and the reliability of life-line systems.

He wrote or co-authored *Conformal Mapping: Methods and Applications* (1991), *Emergencies in Water Delivery* (1979), and *Experiments in Electricity and Magnetism* (1961, with a kit for schools in developing countries). Honors include the Centennial Medal of IEEE, Fellow of IEEE, and Fellow of AAAS. He is a registered professional engineer.

MIKE W. MARTIN received his B.S. (1969, Phi Beta Kappa and Phi Kappa Phi) and M.A. (1972) from the University of Utah, and his Ph.D. (1977) from the University of California, Irvine. Currently he is a professor of philosophy at Chapman University, where he has served as chair of the philosophy department and chair of the Chapman Faculty (in charge of the faculty governance system). He received the Arnold L. and Lois S. Graves Award for Teachers in the Humanities, two fellowships from the National Endowment for the Humanities, grants from the Association of American Colleges, and the 1995 Book Award from the National Society of Fund Raising Executives. Dr. Martin is author of many articles in applied ethics and also eight books, most recently *Meaningful Work: Rethinking Professional Ethics* (2000).

Preface

Technology has a pervasive and profound effect on the contemporary world, and engineering plays a central role in all aspects of technological development. Because of this, it is vital to have an understanding of the ethical implications of engineers' work. Engineers must be aware of their social responsibilities and equip themselves to reflect critically on the moral dilemmas they will confront. Managers must be responsive to the rights of engineers to exercise their consciences responsibly. The public needs to acquire an understanding of the extent and limits of the responsibilities of engineers; it must be prepared to shoulder its own responsibilities where those of the engineers end.

Purpose

Introduction to Engineering Ethics provides the background for discussion of the basic issues in engineering ethics. Emphasis is given to the moral problems engineers face in the corporate setting. It places those issues within a philosophical framework, and it seeks to exhibit both their social importance and their intellectual challenge. The primary goal is to stimulate critical and responsible reflection on the moral issues surrounding engineering practice and to provide the conceptual tools necessary for pursuing those issues.

In large measure, we proceed by clarifying key concepts, sketching alternative views, and providing relevant case study material. Yet, in places, we argue for particular positions that in a subject like ethics can only be controversial. We do so because it better serves our goal of encouraging critical judgment than would a mere digest of others' views. Accordingly, our aim is not to force conviction, but to provoke either reasoned acceptance or reasoned rejection of what we say. We are confident that such reasoning is possible in ethics, and that, through lively and tolerant dialogue, progress can be made in dealing with what at first might seem irresolvable difficulties.

Case studies appear throughout the text, frequently as part of Discussion Topics. Those cases not described in great detail offer the opportunity for practice in literature searches, others that are not analyzed in detail could serve as examples for role-playing in class.

Cases in *engineering* may involve not only engineers but also participants from computer science, mathematics, physics, chemistry, and the biological sciences. This is highlighted in a Taxonomy of Failures (a summary of cases given in appendix A1) by grouping the activities that relate to *engineering, computing, and business ethics.*

Through this grouping we try to remind the reader that engineers work hand in hand with scientists and business professionals and that for practical purposes all their titles could easily be interchanged according to the actual tasks. A broadened perspective on engineering practice comes through studying business ethics and computer ethics.

A cautionary note regarding cases: Most of our case studies are based on secondary sources. Thus each case carries with it an *implied* statement of the sort "If engineer X and company Y did indeed act in the way described, then. . . ." Names of persons and companies are given to make it easier to identify cases, to discuss cases with others, and to search the literature on a case.

It is important to avoid inflexible conclusions regarding persons or organizations based on one or two cases from the past. Persons can and do change with time for the better or the worse, and so can organizations. Thus what was once a proper assessment of a person or an organization may no longer be valid, but the assessment of the actual actions will not have changed.

The engineering profession can learn much from failures, but fear of litigation or possible loss of prestige keeps many companies from openly discussing failures except with a restricted group of confidants. We should also bear in mind the importance of learning from the courageous examples of correction found in the literature.

Courses

Sufficient material is provided for courses entirely devoted to engineering ethics. The book, or chapters of the book, can also be used for a several-week module on engineering ethics within a variety of other courses in engineering, philosophy, business, and the social sciences—courses that typically include such topics as general professional ethics, business ethics, applied philosophical ethics, engineering law, engineering and society, business management, values and technology, engineering design, technology assessment, and safety. Despite the intensity of the engineering curriculum, engineering ethics should enter in several contexts to ensure that students perceive it as a genuine concern of the faculty.

Acknowledgments

Our thinking about engineering ethics has been influenced by too many individuals to list here. We wish to thank especially Robert J. Baum, Michael Davis, Dave Dorchester, Albert Flores, Ed Harris, Deborah G. Johnson, Edwin T. Layton, Jerome Lederer, Heinz C. Luegenbiehl, Steve Nichols, Mike Rabins, Jimmy Smith, Michael S. Pritchard, Harold Sjursen, Carl M. Skooglund, John Stupar, Stephen H. Unger, Pennington Vann, Vivien Weil, and Caroline Whitbeck.

We wish to thank the many authors and publishers who granted us permission to use copyrighted material as acknowledged in the footnotes. We also thank the professional societies that allowed us to print the codes of ethics in the appendix. To promote contact with the professional and student societies representing their respective fields of engineering, the reader is asked to obtain the latest editions of the ethics code and any associated guidelines from the local branch of the respective society or, where such a branch does not exist, via the society's web page.

Special thanks go to Shirley Price for her editing efforts and suggestions, many of which forced us to seek further clarity ourselves.

Roland Schinzinger
Mike W. Martin

McGraw-Hill's Best—Basic Engineering Series and Tools

1

The Profession of Engineering

Engineers create products and processes to improve food production, shelter, communication, transportation, and protection against natural calamities—and, in addition, strive to enhance the convenience and beauty of our everyday lives. They have made possible spectacular human triumphs once only dreamed of in myth and science fiction. A century ago in *From the Earth to the Moon,* Jules Verne imagined American space travelers being launched from Florida, circling the moon, and returning to splash down in the Pacific Ocean. In December of 1968, three astronauts aboard an *Apollo* spacecraft did exactly that. Seven months later, on July 20, 1969, Neil Armstrong took the first human steps on the moon. This extraordinary event was shared with millions of earthbound people watching the live broadcast on television. Engineering had transformed our sense of connection with the cosmos and even fostered dreams of routine space travel for ordinary citizens.

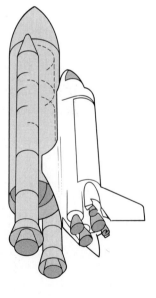

Most technology, however, has double implications: As it creates benefits, it raises new moral challenges. Just as exploration of the moon, followed by equally impressive explorations of planets by unmanned vehicles, will stand as engineering triumphs, so the explosion of space shuttle *Challenger* in 1986 was a tragedy that could have been prevented. We will examine the *Challenger* and other cases of human error, for in considering ethics and engineering alike, we can learn from seeing how things go wrong. Indeed, in a recently discovered manuscript, *Paris in the 20th Century,* Jules Verne himself foresaw long-term problems arising from uncritical applications of technology. But we shall have the opportunity to also bring to mind the many positive dimensions of engineering that so deeply enrich our lives.

This chapter begins with brief characterizations of engineering, ethics, and the goals in studying engineering ethics—a field of study that matured during the last two decades and on which much can be found in the literature now.[1] Next, the idea of engineering as a profession is explored, with an emphasis on the role of professional codes of ethics. Finally, we attend to the corporate setting in which today nearly all engineering is practiced.

Engineering and Moral Complexity

One could, with some exaggeration perhaps, claim that a chance of danger places one in a constant state of crisis. Thus the Chinese combined these characters

danger

chance

to read "crisis" when combined (but also "opportunity")

Engineering: From Concept to Product

Moral values are embedded in the engineering process itself, rather than merely imposed on it by external rules and laws. Likewise, ethical issues arise as a product develops from a mental concept to physical completion. Engineers encounter both moral and technical problems concerning variability in the materials available to them, the quality of work by co-workers at all levels, pressures imposed by time and the whims of the marketplace, and relationships of authority within corporations. The manifold activities of engineers will be described with the help of Figure 1–1, which charts the sequence of tasks that lead from the concept of a product to its design, manufacture, sale, use, and ultimate disposal.

For convenience, several terms are used in broad, generic senses. Thus, the word *product* can apply to a small mass-produced household appliance, an entire communication system, or an oil refinery complex. *Manufacturing* can occur on a factory floor or at a construction site. *Engineers* might appear on the stage singly or in teams, and can be self-employed consultants or entrepreneurs, but mostly they still are employees of large corporations even though the number of engineers working in smaller firms is increasing, as is the frequency at which they change employers. For *corporation,* one may substitute in many cases any structured engineering organization or grouping of engineers, such as a consulting firm or the public works department of a city. The *task* itself might start with a mere concept of a new product, or it may involve improvement of an existing product. Again, the task may call only for a design to be submitted as part of a proposal, or conversely call only for the manufacture of a product according to complete drawings and specifications submitted by another party.

The idea of a brand-new (*de novo*) product is first captured in a conceptual design, which will lead to establishing performance specifications and conducting a preliminary analysis based on the functional relationships among design variables. These activities lead to a more detailed analysis, possibly assisted by computer simulations and physical models or prototypes. The end product of the design task will be detailed specifications and shop drawings for all components.

---

[1] See the selection of sources at the end of this chapter. Specific references appear as footnotes hereafter.

The next major task is manufacturing. It involves scheduling
and carrying out the tasks of purchasing materials and components, fabricating parts and subassemblies, and finally assembling and performance-testing the product.

Selling is next (or delivery, if the product is the result of a
prior contract), and thereafter either the manufacturer's or the
customer's engineers perform installation, personnel training,
maintenance, repair, and ultimately recycling or disposal.

Seldom is the process carried out in such a smooth, continuous
fashion as indicated by the heavy arrows progressing down the
middle of Figure 1–1. Instead of this uninterrupted sequence,

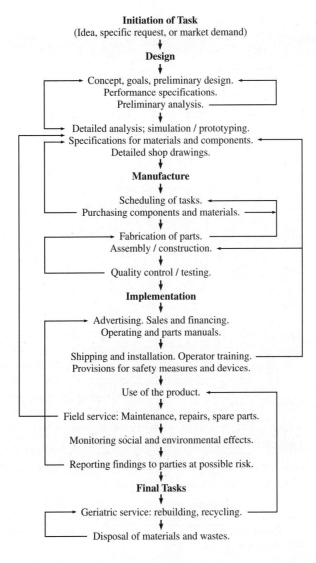

Figure 1–1
Progression of
engineering tasks
(➤ ideal progression,
— typical iterations)

intermediate results during or at the end of each stage frequently require backtracking to make modifications in the design developed thus far. Errors need to be detected and corrected. Alterations may be needed to improve product performance or to meet cost and time constraints. An altogether different, alternative design might have to be considered. In the words of Herbert Simon, "Design is usually the kind of problem solving we call ill-structured . . . you don't start off with a well-defined goal. Nor do you start off with a clear set of alternatives, or perhaps any alternatives at all. Goals and alternatives have to emerge through the design process itself: one of its first tasks is to clarify goals and to begin to generate alternatives."[2]

This results in an iterative process, with some of the possible recursive steps indicated by the thin lines and arrows on either side of Figure 1–1. As shown, engineers are usually forced to stop during an initial attempt at a solution when they hit a snag or think of a better approach. They will then return to an earlier stage with changes in mind. Such reconsiderations of earlier tasks do not necessarily start and end at the same respective stages during subsequent passes through design, manufacture, and implementation. That is because the retracing is governed by the latest findings from current experiments, tempered by the outcome of earlier iterations as well as experience with similar, earlier product designs.

Changes made during one stage will not only affect subsequent stages but may also require an assessment of prior decisions. Requests for design changes while manufacture or construction is in progress must be handled with particular care, otherwise tragic consequences such as the Hyatt-Regency walkway failure illustrated in Figure 1–2 may result. Dealing with this complexity requires close cooperation among the engineers of many different departments and disciplines such as chemical, civil, electrical, industrial, and mechanical engineering. It is not uncommon for engineering organizations to suffer from "silo mentality," which makes engineers disregard or hold in low esteem the work carried out by groups other than their own. It can be difficult to improve a design or even to rectify mistakes under such circumstances. Engineers do well to establish contact with colleagues across such artificial boundaries so that information can be exchanged more freely. Such contacts become especially important when there is a need to tackle morally complex problems.

---

[2] Herbert A. Simon, "What We Know about Learning," *Journal of Engineering Education* (American Society of Engineering Education) 87 (October 1998), pp. 343–48.

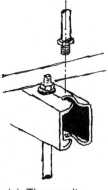

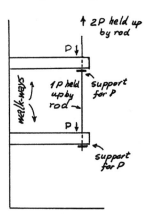

(a) As designed (b) As modified (c) The result

(d) Loads for case (a)

Figure 1–2
The Kansas City Hyatt-Regency walkway collapse. Two walkways—one above the other—along one wall of a large atrium are to be supported by welded box-beams, which in turn are held up along the atrium side by long rods extending from the ceiling. Because of perceived difficulties in implementing design (a), the modification (b) using two shorter rods to replace each long rod was proposed and approved. What is the result? Let the expected load on each box-beam at its atrium end be P (the same on each floor). Then, in design (a), an upper-floor beam would have to support P pounds, as shown in sketch (d), but the design change raised that to $2P$ as shown in (e). This overload caused the box-beam/rod/nut supports on the upper floor to fail as shown in (c). In turn, the upper and lower walkways collapsed, causing a final death toll of 114 with 200 injured. Later it was found that the design change had been stamped "approved" but not checked. (For more, see M. Levy and M. Salvadori, *Why Buildings Fall Down* [Norton & Co., 1992].)

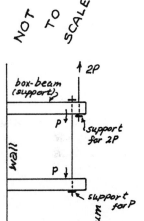

(e) Loads for case (b)

From Problem Solving to Decision Making

To repeat, engineering is not a successive problem solving along a straightforward progression of isolated tasks. Instead, it involves a trial-and-error process with backtracking based on decisions made after examining results obtained along the way. Note also that the design iterations resemble feedback loops.[3] But like any well-functioning feedback control system, our model of engineering is not complete until we take into account natural and social environments that affect the product and the people using it. Let us therefore revisit the engineering tasks, this time as listed in Table 1–1 along with examples of problems that might arise.

_____

[3] Roland Schinzinger, *"Ethics* on the Feedback Loop," *Control Engineering Practice* 6 (1998), pp. 239–45. See also Harris, Pritchard, and Rabins, *Engineering Ethics;* and Whitbeck, *Ethics in Engineering Practice and Research,* for use of "feedback" in resolving ethical problems.

Table 1–1 Engineering tasks and possible problems

Tasks	A selection of possible problems
Conceptual design	Blind to new concepts. Violation of patents or trade secrets. Product to be used illegally.
Goals; performance specifications	Unrealistic assumptions. Design depends on unavailable or untested materials.
Preliminary analysis	Uneven: Overly detailed in designer's area of expertise, marginal elsewhere.
Detailed analysis	Uncritical use of handbook data and computer programs based on unidentified methodologies.
Simulation, prototyping	Testing of prototype done only under most favorable conditions or not completed.
Design specifications	Too tight for adjustments during manufacture and use. Design changes not carefully checked.
Scheduling of tasks	Promise of unrealistic completion date based on insufficient allowance for unexpected events.
Purchasing	Specifications written to favor one vendor. Bribes, kickbacks. Inadequate testing of purchased parts.
Fabrication of parts	Variable quality of materials and workmanship. Bogus materials and components not detected.
Assembly/construction	Workplace safety. Disregard of repetitive-motion stress on workers. Poor control of toxic wastes.
Quality control/testing	Not independent, but controlled by production manager. Hence, tests rushed or results falsified.
Advertising and sales	False advertising (availability, quality). Product oversold beyond client's needs or means.
Shipping, installation, training	Product too large to ship by land. Installation and training subcontracted out, inadequately supervised.
Safety measurers and devices	Reliance on overly complex, failure-prone safety devices. Lack of a simple "safety exit."
Use	Used inappropriately or for illegal applications. Overloaded. Operations manuals not ready.
Maintenance, parts, repairs	Inadequate supply of spare parts. Hesitation to recall the product when found to be faulty.
Monitoring effects of product	No formal procedure for following life cycle of product, its effects on society and environment.
Recycling/disposal	Lack of attention to ultimate dismantling, disposal of product, public notification of hazards.

The grab bag of problems in Table 1–1 may arise from short-comings on the part of engineers, their supervisors, vendors, or the operators of the product. The underlying causes can have different forms:

a. *Lack of vision,* which in the form of tunnel vision biased toward traditional pursuits overlooks suitable alternatives, and in the form of *groupthink* (a term coined by Irving Janis) promotes acceptance at the expense of critical thinking.

b. *Incompetence* among engineers carrying out technical tasks.

c. A *lack of time* or *lack of proper materials,* both ascribable to poor management.

d. A *silo mentality* that keeps information compartmentalized rather than shared across different departments.

e. The notion that there are safety engineers *somewhere down the line* to catch potential problems.

f. *Improper use* or *disposal of the product* by an unwary owner or user.

g. *Dishonesty* in any activity shown in Figure 1–1.

This list is not complete, but it serves to hint at the range of problems that can generate moral difficulties for engineers. As may be expected, the problems encountered depend very much on the type of product, the range of tasks undertaken, the organizational structure, and market conditions.

Engineers need foresight and caution (concepts that are combined in the German word *Vorsicht*). They need the ability to construct scenarios to help them imagine who may be affected indirectly by their products and by their decisions, in good or harmful ways. As Figure 1–3 indicates, there are strands linking them personally or through their work to clients, the community, a host of organizations representing their industry, their profession, the government, and even the natural environment.

Of special importance are bonds to their families, which brings us to the question of priorities. If an engineer faces a problem that can be solved in several ways that affect different parties differently, who will be given priority? Should it be the engineer's family? After all, some decisions could threaten the engineer's job security, either through dismissal because the employer is displeased or because the employer goes bankrupt (and must dismiss others as well). The problem of conflicting priorities and loyalties is one we will come across again and again. It is at the heart of many of the difficulties engineers face as they make their important decisions.

Another often overlooked but critical set of decisions arises as products and systems perform more functions, or do so more rapidly. Apparently small technical malfunctions can produce massive failures with severe consequences for many people. Never before has quality been so important in preventing harm, and never has safety required so much attention. Yet, when safety mechanisms themselves become automated (and as a consequence

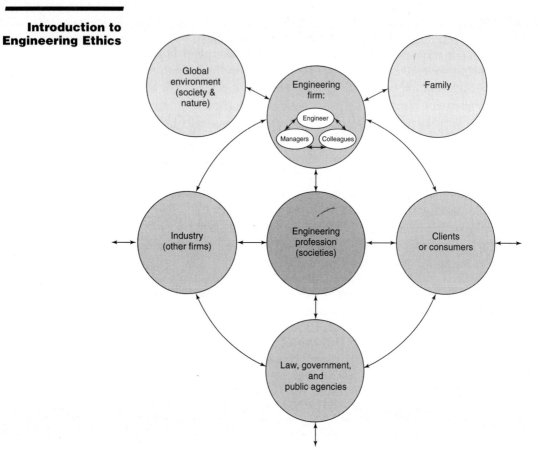

Figure 1–3
Intertwined responsibilities

complicated and frequently error prone), there is no substitute for adding an uncomplicated "safe exit" for when things go awry.

Defining Engineering Ethics

The word *ethics* has several distinct although related meanings, and hence so does the expression *engineering ethics*. After distinguishing three main usages, we will rely on the context to indicate which meaning is intended.

In the sense used most often in this book, *ethics* refers to an *area of inquiry.* It is the activity of understanding moral values, resolving moral issues, and justifying moral judgments. It is also the discipline or area of study resulting from that activity. *Engineering ethics,* accordingly, *is the study of the moral values, issues, and decisions involved in engineering practice.* The moral values take many forms, including responsibilities, ideals, character traits, social policies, and relationships desirable for individuals and corporations engaged in technological development.

In a second sense, the word *ethics* refers to the particular _beliefs or attitudes_ concerning morality that are endorsed by specific groups or individuals. Using this sense, engineering ethics consists of the requirements specified in the currently accepted engineering codes of ethics, such as those included at the end of this book. Alternatively, engineering ethics might refer to the actual conduct of individuals who are engineers.

In a third sense, the word *ethics* and its grammatical variants are synonyms for "morally correct" or justified. In this usage, engineering ethics amounts to the set of justified moral principles of obligation, rights, and ideals that ought to be endorsed, as they apply generally and to engineering in particular, by those engaged in engineering. Clarifying such principles and applying them to concrete situations is the central goal of engineering ethics as an area of study.

What Is Morality? what you learn

Engineering ethics studies moral values in engineering, but what are moral values? What is morality?

One suggestion given in dictionaries is that morality concerns right and wrong, good and bad, the rules that ought to be followed. This definition is incomplete, however, because there are nonmoral as well as moral usages of these words. Thus, in order to start a car, a person ought to put the key in the ignition—that is the right thing to do. Again, it is good to get up early on a pleasant spring day in order to experience the value (beauty) of a sunrise. These are not moral judgments, which instead are about what is *morally* right or wrong, morally good or bad, or morally valuable or harmful, and morally ought or ought not to be done. Because *morally* is a grammatical variation of the word we are trying to define, we are caught in an unhelpful circle of definitions.

It turns out that morality is not easily encapsulated in a simple definition. In the next chapter, we will discuss some ethical theories that provide comprehensive perspectives about morality. For now, let us say that morality is about reasons centered in respect for other people as well as for ourselves, reasons that involve caring for their good as well as our own. Moral reasons, for instance, involve respecting persons by being fair and just with them, respecting their rights, keeping promises, avoiding unnecessary offense and pain to them, and avoiding cheating and dishonesty. They also concern caring for others by sometimes being willing to help them (especially when they are in distress), showing gratitude for favors, and empathizing with their suffering. In addition, moral reasons extend to concern for minimizing suffering to animals and damage to the environment.

Illustrative Cases

The following examples hint at a few types of moral issues in engineering.

1. An inspector discovered faulty construction equipment and applied a violation tag, preventing its continued use. The inspector's supervisor, a construction manager, viewed the case as a minor infraction of safety regulations and ordered the tag removed so the project would not be delayed. The inspector objected and was threatened with disciplinary action. The continued use of the equipment led to the death of a worker on a tunnel project.

2. An electric utility company applied for a permit to operate a nuclear power plant. The licensing agency was interested in knowing what emergency measures had been established for human safety in case of reactor malfunction. The utility engineers described the alarm system and arrangements with local hospitals for treatment. They did not emphasize that these measures applied to plant personnel only and that they had no plans for the surrounding population. "That is someone else's responsibility, but we don't know whose," they answered upon being questioned about this omission.

3. A chemical plant dumped wastes in a landfill. Hazardous substances found their way into the underground water table. The plant's engineers were aware of the situation but did not change the disposal method because their competitors did it the same cheap way, and no law explicitly forbade the practice. Plant supervisors told the engineers it was the responsibility of the local government to identify any problems.

4. The ABC company began selling its latest high-tech product before it had been fully checked out in beta tests, that is, used on real applications by a group of knowledgeable users. It was not really ready for distribution, but clients were already lured to this product by glossy advertising designed to win the market by being first to capture clients' attention.

5. An industrial engineer working for the Air Force had repeatedly reported cost overruns incurred on contracts for the development of airplanes. That had already made him unpopular with his supervisors when, in 1968, he was asked by a Senate committee to testify. Telling more than his supervisors wanted him to turned into a nightmare for him and his family. Eventually he was fired, only to be reinstated by a court order years later. This case

of Ernest Fitzgerald will be taken up in Chapter 5 in connection with whistle-blowing.

6.   Therac-25, a linear electron accelerator for therapeutic use was built as a dual-mode system that could produce either X-rays or electron beams. It had been in successful use for some time, but every now and then, some patients received high overdoses, resulting in painful after-effects and several deaths. One patient on a repeat visit experienced great pain, but the remotely located operator was unaware of any problem because of lack of communication between them: The intercom was broken and the video monitor had been unplugged. There also was no way for the patient to exit the examination chamber without help from the outside, and hence the hospital was partly at fault. Upon cursory examination of the machine, the manufacturer insisted that the computerized and automatic control system could not possibly have malfunctioned and that no one should spread unproven and potentially libelous information about the design. It was the painstaking, day-and-night effort of the hospital's physicist that finally traced the problem to a software error introduced by the manufacturer's efforts to make the machine more user-friendly.[4]

These examples show how ethical problems arise most often when there are differences of judgment or expectations as to what constitutes the true state of affairs or a proper course of action. They also raise a number of general moral questions. To what extent should an employer's or supervisor's directives be the authoritative guide to an engineer's conduct? What does one do when there are differences of judgment? Is it fair to be expected to put one's job on the line? Should one always follow the law to the letter? Is an engineer to do no more than what the specifications say, even if there are problems more serious than those initially anticipated? How far does an engineer's responsibility extend into the realm of anticipating and influencing the social impact of the projects in which he or she participates?

The case of the inspector told to ignore rules has numerous variations: in the testing of prototypes or finished products in a factory, in the handling of financial audits, or in the response to unsolicited reports of dangerous situations, to name but a few. The individual who has the courage to take action contrary to a supervisor's directives can suffer severe consequences and not be vindicated until much later, as the whistle-blowing case shows.

The nuclear power plant situation demonstrates several ways in which a product's ultimate operation and related consequences

---

[4] N. G. Leveson and C. Turner, "An Investigation of the Therac-25 Accidents," *Computer,* IEEE, July 1993, pp. 18–41; R. Schinzinger, "Ethics on the Feedback Loop."

may be overlooked during the planning process. Lack of coordination in planning for its effects on everyone, even the uninvolved bystander, can lead to catastrophic results; limiting one's attention only to the narrow specifications can result in the delivery of a product that satisfies the contract but does not serve the needs of the customer or the public in the long run.

Electron beam accelerators are potent radiation devices and need to be designed, built, and operated with great care. As the sixth case shows, attention to safety is a must, and even such apparently insignificant devices as intercoms and video survey cameras must be in top operating condition in critical places. More telling in this case is the determined effort at another hospital to get at the roots of a life-threatening problem, rather than being content to blame "operator error" as is so often the case.

The questionable conduct of the chemical plant and the ABC company falls into the category of "ethics cases" that traditionally have occupied the ethics review boards of professional societies. Prescriptive codes of conduct can be established for such infractions of common decency, and for that reason those instances constitute the majority of cases in which action has been taken. This has encouraged the identification of engineering ethics with a finite set of specific maxims and regulations designed to assure moral conduct. Yet, the moral problems in engineering ethics are not always manageable in a straightforward manner, neither in preventing nor in adjudicating them.

Moral Dilemmas and Related Issues

Moral reasons are many and varied, and they frequently come into conflict, creating moral dilemmas. *Moral dilemmas* are situations in which two or more moral obligations, duties, rights, goods, or ideals come into conflict with one another. It is also possible for one moral principle to have two or more incompatible applications in a given situation. Each of the preceding cases involved a dilemma, some with obvious solutions and some with the greater uncertainty expressed in everyday language by the term *dilemma.*

Because moral principles can conflict, it is often difficult or impossible to formulate rules that are *absolute,* that is, never have a justified exception. Even such basic principles as Do not lie, Do not steal, and Do not kill have some permissible exceptions when they conflict with more pressing moral duties. Most moral principles are *prima facie,* in a technical philosophical sense: They can have justified exceptions when they conflict with other rules that are more important in the situation. Resolving moral dilemmas involves good moral judgment in weighing conflicting moral reasons, but frequently it involves several related tasks: conceptual clarification, factual inquiries, and resolution of interpersonal disagreements.

Conceptual clarification is the elucidation of moral ideas and morally relevant notions. We have already been engaged in conceptual clarification when we defined basic ideas like "engineering ethics" and "morality." As we proceed, we seek to clarify many others, such as "honesty," "bribes," and "whistle-blowing." In general, moral ideas, like most ideas in everyday life, contain areas of vagueness (lack of a clear meaning) and ambiguity (two or more meanings) that need to be dealt with.

Factual inquiries are inquiries into the facts relevant to resolving particular moral issues. These are engineering, scientific, financial, or legal matters, as they are brought to bear on resolving moral dilemmas. Usually, the controversy is interpersonal, as in a disagreement between two persons or groups about what the pertinent facts are, although of course a controversy can also be internal to an individual who is uncertain about what should be believed.

Interpersonal disagreements are controversies among persons or groups about how to understand and resolve moral dilemmas. Ideally, rational dialogue brings people to a rough and workable consensus, either through mutually enriching perspective or through mutual compromises, but the world is rarely ideal and conflicts abound.

Drawing by Dana Fradon; © 1975 The New Yorker Magazine, Inc.

'Miss Dugan, will you send someone in here who can distinguish right from wrong?'

Steps in Confronting Moral Dilemmas

There are several steps in approaching moral dilemmas. The steps are distinct, even though they are interrelated and can often be taken jointly.

1. Identify the relevant moral factors and reasons. What are the conflicting responsibilities, competing rights, clashing ideals, and goods and bads involved?

2. Gather all available facts that are pertinent to the moral factors involved.

3. If possible, rank the moral considerations in order of importance as they apply to the situation. Sometimes this is not possible, and the goal is to find a way to meet equally urgent responsibilities and to promote equally important ideals.

4. Consider alternative courses of action as ways of resolving the dilemma, tracing the full implications of each. Typically this involves making factual inquiries.

5. Talk with colleagues (or friends or other students), seeking their suggestions and alternative perspectives on the dilemma.

6. Arrive at a carefully reasoned judgment by weighing all the relevant moral factors and reasons in light of the facts.

Why Study Engineering Ethics?

Let us conclude by noting the main goals in studying engineering ethics. The briefest answer to why engineering ethics should be studied is that it is *important,* both in preventing grave consequences of faulty ethical reasoning and in giving meaning to engineers' endeavors, and it is *complex,* in ways that cannot be understood through casual observations. As the preceding discussion has indicated, engineering and morality are each complicated, and their conjunction doubles the complexity. But more fully, what should be the specific goals in studying engineering ethics?

In our view, the direct aim is to increase the ability to deal effectively with moral complexity in engineering. Accordingly, the study of engineering ethics aims at empowering individuals to reason more clearly and carefully concerning moral questions, rather than to inculcate any particular beliefs. To invoke a term widely used in ethics, the unifying goal is to increase moral autonomy.

Autonomy literally means "self-determining" or "independent." But not just any kind of independent reflection about ethics amounts to moral autonomy. Moral autonomy can be viewed as the skill and habit of thinking rationally about ethical issues on the basis of moral concern. This foundation of moral concern, or general responsiveness to moral values, derives primarily from the training we receive as children in being sensitive to the needs and rights of others, as well as of ourselves. Where such training is absent, as it often is with abused or neglected children, the tragic result can be an adult sociopath capable of murdering without compunction. Sociopaths, who by definition lack a sense of moral concern and guilt, are never morally autonomous—no matter how "independent" their intellectual reasoning about ethics may be.

Improving the ability to reflect carefully on moral issues can be accomplished by improving various practical skills that will help produce autonomous thought about moral issues. As related to engineering ethics, these skills include the following:

1. Proficiency in recognizing moral problems and issues in engineering.
2. Skill in comprehending, clarifying, and assessing critically arguments on opposing sides of moral issues.
3. The ability to form consistent and comprehensive viewpoints based on consideration of relevant facts.
4. Imaginative awareness of alternative responses to the issues and receptivity to creative solutions for practical difficulties.
5. Increased precision in the use of a common ethical language, a skill needed to express and support one's moral views adequately to others.
6. Sensitivity to genuine difficulties and subtleties. This includes a willingness to undergo and tolerate some uncertainty in making troublesome moral judgments or decisions.
7. An awakened sense of the importance of integrating one's professional life and personal convictions—that is, the importance of maintaining one's moral integrity.
8. Enriched appreciation of both the possibilities of using rational dialogue in resolving moral conflicts and of the need for tolerance of differences in perspective among morally reasonable people.

Discussion Topics

The following questions all center on moral issues that emerged with the rapid development of the Internet during the 1990s.[5] As the most powerful communication technology ever developed, the Internet is a wellspring of new ways to be in contact with other

---

[5] M. David Ermann, Mary B. Williams, and Claudio Gutierrez, eds., *Computers, Ethics, and Society* (New York: Oxford University Press, 1990); Deborah G. Johnson, *Computer Ethics,* 2nd ed. (Englewood Cliffs, NJ: Prentice Hall, 1994); Richard A. Spinello, *Case Studies in Information and Computer Ethics* (Upper Saddle River, NJ: Prentice Hall, 1997).

people and sources of information. It has also created greater convenience in ordering consumer items, paying bills, and trading stocks and bonds. Like other major new technologies, however, it also has raised a host of new issues in information ethics (we can call it "informethics" for short).

With respect to each of the following topics, (a) state what you see as any moral dilemmas (that is, situations where two or more moral reasons conflict); (b) identify any factual inquiries you think might be needed in making a reliable judgment about the case; (c) identify any ideas or concepts involved in dealing with the moral issues that it would be useful to clarify; (d) identify sources of conflict among individuals and groups; (e) present and defend your view about how best to resolve the moral issue.

1. Free speech and bigotry: The use of the Internet by hate groups to spread racist, sexist, and sometimes violent agendas.

2. Speech and exploitation: Pornography, and specifically child pornography.

3. Bulletin boards: Should bulletin board and web-site operators be held liable for failing to filter illegal forms of verbal assaults, even if they are forced to buy liability insurance and thereby raise the costs of creating information sites?

4. Privacy and security, especially concerning financial information.

5. Gambling on the Internet ("techno-gambling" using "cyberspace casinos"). What about trading on the NASDAQ stock exchange and day trading—does that approach gambling? (Related issues: Does the "house" have certain advantages unknown to the casual participant? Can rapid churning of shares and surfing the yield curve of foreign currencies lead to market instabilities? Should trading of exceptionally volatile stocks be halted temporarily, and who will decide that?)

6. False identities: Creating and communicating to others a false persona on the Internet.

7. Property: Have I stolen your property if I publish your essay on the Internet, under your name but without your permission?

8. "Spamming": Should I be allowed to overload your system with endless advertisements?

9. In 1994, charges of computer fraud were filed against David LaMacchia, a 20-year-old student at Massachusetts Institute of Technology (MIT).[6] LaMacchia operated a computer bulletin board that allowed users to download copyrighted software.

---

[6] Spinello, *Case Studies in Information and Computer Ethics,* pp. 127–29. Consult also the Web at www.onlineethics.org.

LaMacchia did not personally profit from the "service" he provided, and apparently his motives were based on his convictions that software should be more freely available to everyone in a free society.

Engineering ethics is a branch of professional ethics, that is, the study of moral values and issues in the professions. Professional organizations have addressed the complexity of moral issues in their fields by developing codes of ethics. Those codes have great importance as an expression of the profession's collective commitment to ethics, even though codes are not the full substance of professional ethics. They also emerge naturally from the very structure and functions of professions.

What Are Professions?

In a wide sense, a profession is any occupation that provides a means to earn a living. In the sense intended here, however, professions are those forms of work involving advanced expertise, self-regulation, and concerted service to the public good.[7]

1. *Advanced Expertise.* Professions require sophisticated skills ("knowing-how") and theoretical knowledge ("knowing-that") in exercising judgment that is not entirely routine or susceptible to mechanization. Preparation to engage in the work requires extensive formal education, including technical studies in one or more areas of systematic knowledge as well as broader humanistic studies. Generally, continuing education and updating knowledge are also required.

2. *Self-Regulation.* Well-established societies of professionals are allowed by the public to play a major role in setting standards for admission to the profession, drafting codes of ethics, enforcing standards of conduct, and representing the profession before the public and the government.

3. *Public Good.* The occupation serves some important aspect of the public good. For example, medicine is directed toward promoting health, law toward protecting the public's legal rights, and engineering toward technological solutions to problems concerning the public's well-being, safety, and health.

Roles of Codes of Ethics

Professional codes of ethics consist primarily of principles of responsibility that delineate how to promote the public good. As

---

[7] Michael Bayles, *Professional Ethics,* 2nd ed. (Belmont, CA: Wadsworth, 1989); Joan C. Callahan, ed., *Ethical Issues in Professional Life* (New York: Oxford University, 1988).

such, codes provide guidance and support for responsible engineers, establish shared minimum standards, and play additional important roles.

Shared Standards

The great diversity of moral views makes it essential that professions establish explicit standards. Even when the standards are less than ideal, the mere fact of shared standards throughout the profession has value. In this way, the public is assured of a minimum standard of excellence on which it can depend, and professions are provided a fair playing field allowing competition.

Support

Codes give positive support to those seeking to act ethically. A publicly proclaimed code allows an engineer, under pressure to act unethically, to say: "I am bound by the code of ethics of my profession, which states that . . ." This by itself gives engineers some group backing in taking stands on moral issues. Moreover, codes can potentially serve as legal support for engineers when criticized for living up to work-related professional obligations.

Guidance

Codes provide a positive stimulus for ethical conduct and helpful guidance concerning the main obligations of engineers. Since codes should be brief to be effective, they offer mostly general guidance. More specific directions may be given in supplementary statements or guidelines, which tell how to apply the code. Further specificity may also be attained by the interpretation of codes. This is done for engineers by the National Society of Professional Engineers (NSPE). It has established a Board of Ethical Review that applies the Society's code to specific cases. It publishes the results in *Professional Engineer* and in periodic volumes entitled *NSPE Opinions of the Board of Ethical Review.*

Inspiration

Because codes express a profession's collective commitment to ethics, they provide a positive stimulus (motivation) for ethical conduct. In doing so, they often use language with positive overtones that introduces a large element of vagueness, as in phrases like "safeguard the public safety, health, and welfare," a vagueness that may lessen the code's ability to give concrete guidance even though it sets forth important general ideals.

Education and Mutual Understanding

Codes can be used by professional societies and in the classroom to prompt discussion and reflection on moral issues. Widely circulated and officially approved by professional societies, codes encourage a shared understanding among professionals, the public, and government organizations concerning the moral responsibilities of engineers.

Deterrence and Discipline

Codes can also serve as the formal basis for investigating unethical conduct. Where such investigation is possible, a deterrent for immoral behavior is thereby provided. Such an investigation generally requires paralegal proceedings designed to get at the truth about a given charge without violating the personal rights of those being investigated. Unlike the American Bar Association and some other professional groups, engineering societies cannot by themselves revoke the right to practice engineering in the United States. Yet some professional societies do suspend or expel members whose professional conduct has been proven unethical, and this alone can be a powerful sanction when combined with the loss of respect from colleagues and the local community that such action is bound to produce.

Contributing to the Profession's Image

Codes can present a positive image to the public of an ethically committed profession. Where the image is warranted, it can help engineers more effectively serve the public. It can also win greater powers of self-regulation for the profession itself, while lessening the demand for more government regulation.

Abuse of Codes

When codes are not taken seriously within a profession, they amount to a kind of window dressing that ultimately increases public cynicism about the profession. Worse, codes occasionally stifle dissent within the profession and are abused in other ways.

Probably the worst abuse of engineering codes occurs when honest moral effort on the part of individual engineers is restricted by an attempt to preserve the profession's public image and protect the status quo. Preoccupation with keeping a shiny public image may silence healthy dialogue and criticism. And an excessive interest in protecting the status quo may lead to a distrust of the engineering profession on the part of both government and the public. The best way to increase trust is by encouraging and aiding engineers to speak freely and responsibly about public safety and well-being. And this includes a tolerance for criticisms of the codes themselves, rather than allowing codes to become "sacred documents" that have to be accepted uncritically.

On occasion, this abuse has positively discouraged moral conduct and caused serious harm to those seeking to serve the public. In 1932, for example, two engineers were expelled from ASCE for violating a section of its code forbidding public remarks critical of other engineers. Yet the actions of those engineers were essential in uncovering a major bribery scandal related to the construction of a dam for Los Angeles County.[8]

Moreover, codes have sometimes placed unwarranted "restraints of commerce" on business dealings to benefit those within the profession. Obviously there is disagreement about which, if any, entries function in these ways. Consider the following entry in the pre-1979 versions of the NSPE code: The engineer "shall not solicit or submit engineering proposals on the basis of competitive bidding." This prohibition was felt by the NSPE to best protect the public safety by discouraging cheap engineering proposals that might slight safety costs in order to win a contract. The U.S. Supreme Court ruled, however, that it mostly served the self-interest of established engineering firms and actually hurt the public by preventing the lower prices that might result from greater competition [*The National Society of Professional Engineers v. the United States,* (April 25, 1978)].

Limitations of Codes

Codes are no substitute for individual responsibility in grappling with concrete dilemmas. For instance, most codes are restricted to general wording, and hence inevitably contain substantial areas of vagueness. Thus, they may not be able to straightforwardly address all situations. At the same time, vague wording may be the only way new technical developments and shifting social and organizational structures can be accommodated.

Other uncertainties can arise when different entries in codes come into conflict with each other. Usually codes provide no guidance as to which entry should have priority in those cases, thereby creating moral dilemmas.

A further limitation of codes results from their proliferation. Andrew Oldenquist (a philosopher) and Edward Slowter (an engineer and former NSPE president) point out how the existence of separate codes for different professional engineering societies can give members the feeling that ethical conduct is more relative and variable than it actually is.[9] But Oldenquist and Slowter have also demonstrated the substantial agreement to be found among the various engineering codes, urging that the

---

[8] Edwin T. Layton, "Engineering Ethics and the Public Interest: A Historical View," in *Ethical Problems in Engineering,* vol. 1, ed. Albert Flores (Troy, NY: Rensselaer Polytechnic Institute, 1980), pp. 26–29.

[9] Andrew G. Oldenquist and Edward E. Slowter, "Proposed: A Single Code of Ethics for All Engineers," *Professional Engineer* 49 (May 1979), pp. 8–11.

time has come for adoption of a unified code. Indeed, currently attempts are being undertaken in that direction by umbrella organizations of engineering, such as the American Association of Engineering Societies (AAES) and the Accreditation Board for Engineering and Technology (ABET). Already, the National Society of Professional Engineers (NSPE) provides a unifying code for individuals who are registered professional engineers. In 1995, the North American Free Trade Agreement included a section titled "Principles of Ethical Conduct in Engineering Practice" prepared by U.S., Canadian, and Mexican engineers.

Most important, codes should not be seen as the final moral authority for professional conduct.[10] Codes can be flawed. Again, a possible example is the former ban in engineering codes on competitive bidding. Codes, after all, represent a compromise between differing judgments, sometimes developed amid heated committee disagreements. As such, they have a great "signpost" value in suggesting paths through what can be a bewildering terrain of moral decisions. But equally as such they should never be treated as sacred canon in silencing healthy moral debate.

This limitation of codes connects with a wider issue about whether professional groups or entire societies can create morally authoritative sets of standards for themselves, or whether group standards are always open to moral scrutiny in light of wider values. This is the complex issue of ethical relativism, which will receive further comment in Chapter 2.

To conclude, any set of conventions, whether codes of ethics or actual conduct, should be open to scrutiny in light of wider values. At the same time, professional codes should be taken very seriously. They express the good judgment of many morally concerned individuals, the collective wisdom of a profession at a given time. Change over time is apparent if one compares how principles of professional ethics were stated in the early and late 1900s. Consider the two contrasting examples that appear boxed on the next page. It will be noted how an earlier emphasis on upholding the reputation of engineering has given way to an emphasis on upholding the public good. Certainly codes are the starting place for an inquiry into professional ethics; they establish a framework for dialogue about moral issues; and, more often than not, they cast powerful light on the dilemmas confronting engineers.

---

[10] John Ladd, "The Quest for a Code of Professional Ethics," in Rosemary Chalk, Mark S. Frankel, and Sallie B. Chafer (eds.), *AAAS Professional Ethics Project* (Washington, D.C.: American Association for the Advancement of Science, 1980), pp. 154–59. For an opposing view that heavily emphasizes the importance of codes, see Michael Davis, *Thinking Like an Engineer* (New York: Oxford, 1998).

An Early Statement of Principles

A. General Principles*

1. In all of his relations the engineer should be guided by the highest principles of honor.

2. It is the duty of the engineer to satisfy himself to the best of his ability that the enterprises with which he becomes identified are of legitimate character. If after becoming associated with an enterprise he finds it to be of questionable character, he should sever his connection with it as soon as practicable.

*) From *Code of Principles of Professional Conduct*, American Institute of Electrical Engineer (AIEE),1912. This code appears in full in the Appendix of this book.

A Current Statement of Principles

The Fundamental Principles*

Engineers uphold and advance the integrity, honor, and dignity of the engineering profession by:

I. using their knowledge and skill for the enhancement of human welfare;

II. being honest and impartial, and serving with fidelity the public, their employers and clients;

III. striving to increase the competence and prestige of the engineering profession;

IV. supporting the professional and technical societies of their disciplines.

*) From *Code of Ethics of Engineers*, Accreditation Board for Engineering and Technology (ABET), 1977. These principles (sometimes without item IV) can also be found in the codes of other societies. See Appendix.

Discussion Topics

1. Do the following definitions, or partial definitions, of professionalism express something important, or do they express unwarranted views?

   (a) "Professionalism implies a certain set of attitudes. A professional analyzes problems from a base of knowledge in a specific area, in a manner which is objective and independent of self-interest and directed toward the best interests of his client. In fact, the professional's task is to know what is best for his client even if his client does not know himself."[11]

   (b) "So long as the individual is looked upon as an employee rather than as a free artisan, to that extent there is no professional status."[12]

---

[11] Lawrence Storch, "Attracting Young Engineers to the Professional Society," *Professional Engineer* 41 (May 1971), p. 3.

[12] Robert L. Whitelaw, "The Professional Status of the American Engineer: A Bill of Rights," *Professional Engineer* 45 (August 1975), pp. 37–38.

(c) "A truly professional man will go beyond the call to duty. He will assume his just share of the responsibility to use his special knowledge to make his community, his state, and his nation a better place in which to live. He will give freely of his time, his energy, and his worldly goods to assist his fellow man and promote the welfare of his community. He will assume his full share of civic responsibility."[13]

2. Disputes arise over how a person becomes or should become a member of an accepted profession. Such disputes often occur in engineering. Each of the following has been proposed as a criterion for being a "professional engineer" in the United States. Assess these definitions to determine which, if any, captures what you think should be part of the meaning of *engineers.*

(a) Earning a bachelor's degree in engineering at a school approved by the Accreditation Board for Engineering and Technology. (If applied in retrospect, this would rule out Leonardo da Vinci and Thomas Edison.)

(b) Performing work commonly recognized as what engineers do. (This rules out many engineers who have become full-time managers, but embraces some people who do not hold engineering degrees.)

(c) Being officially registered and licensed as a Professional Engineer (PE). Becoming registered typically includes (1) passing the Engineer-in-Training Examination or Professional Engineer Associate Examination shortly before or after graduation from an engineering school, (2) working four to five years at responsible engineering, (3) passing a professional examination, and (4) paying the requisite registration fees. (Only those engineers whose work directly affects public safety and who sign official documents such as drawings for buildings are required to be registered as PEs. Engineers who practice in manufacturing or teach at engineering schools are exempt. Nevertheless, many acquire their PE licenses out of respect for the profession or for prestige.)

(d) Acting in morally responsible ways while practicing engineering. The standards for responsible conduct might be those specified in engineering codes of ethics or an even fuller set of valid standards. (This rules out scoundrels, no matter how creative they may be in the practice of engineering.)

---

[13] Harry C. Simrall, "The Civic Responsibility of the Professional Engineer," *The American Engineer,* May 1963, p. 39.

3. Examine the codes of ethics at the back of this book. Discuss how they apply to the six illustrative cases discussed earlier in this chapter.

4. Do the codes of ethics at the back of this book adequately tell us what should be done in the following case? If not, can they be amended so as to resolve the issues? How does this case differ from the use of fingerprint checking machines in similar settings? Or iris scanners that are being developed?

A team of engineers and biomedical computer scientists develops a system for identifying people from a distance of up to 200 meters. A short tube attached to a sophisticated receiver and computer, and aimed at a person's head, reads the individual's unique pattern of brain waves when standard words are spoken. The team patents the invention and forms a company to manufacture and sell it. The device is an immediate success within the banking industry. It is used to secretly verify the identification of customers at tellers' windows. The scientists and engineers, however, disavow any responsibility for such uses of the device without customer notification or consent. They contend that the companies that buy the product are responsible for its use. They also refuse to be involved in notifying public representatives about the product's availability and the way it is being used. Does employing the device without customer awareness violate privacy? Do the engineers and scientists perhaps have a moral obligation to market the product with suggested guidelines for its ethical use? Should they be involved in public discussions about permissible ways of using it? [14]

Corporations and Responsibility

From its inception as a profession (distinct from a craft), much of engineering has been embedded in corporations. That is due to the nature of engineering, both in its goal of producing economical and safe products for the marketplace and in its usual requirement that many individuals work together.

Historian Edwin T. Layton identified two main stages in the development of engineering as a profession during the nineteenth century.[15] First, the growth of public resources during the first half of the century made possible the extensive building of railroads, canals, and other large projects that only large technological organizations could undertake. Second, from 1880

---

[14] Donn B. Parker, *Ethical Conflicts in Computer Science and Technology* (Arlington, VA: AFIPS Press, 1979), pp. 126–27.

[15] Edwin T. Layton Jr., *The Revolt of the Engineers: Social Responsibility and the American Engineering Profession* (Baltimore: Johns Hopkins University Press, 1986).

to 1920, the demand for engineers exploded, increasing their ranks 20 times over. Along with this increase came a demand for science- and mathematics-based training, as engineering schools began to multiply.

Traditionally, professions and professionals were thought of as being independent, with the paradigm being the country doctor or attorney; hence, engineers employed by companies and agencies were frequently not recognized as full-blown professionals. Today, the dramatic trend is for even the more traditional professions, such as medicine and law, to function in corporate settings. Hence, many more professionals must now confront the tensions between business and professional interests with which most engineers have always had to deal. In the main, those dual interests are complementary and mutually reinforcing, and they are both supported by sound ethical practices. This is a theme we will emphasize in this book: Good engineering, good business, and good ethics work together in the long run.

An Ethical Corporate Climate

An ethical climate is a working environment conducive to morally responsible conduct. Within corporations, it is produced by a combination of formal organization and policies, informal traditions and practices, and personal attitudes and commitments. Engineers can make a vital contribution to such a climate, especially as they move into technical management and then more general management positions.

Professionalism in engineering would be threatened at every turn in a corporation devoted exclusively to powerful egos. Sociologist Robert Jackall describes several such corporations in his book, *Moral Mazes,* as organizations that reduce (and distort) corporate values to merely following orders: "What is right in the corporation is what the guy above you wants from you. That's what morality is in the corporation."[16] Jackall describes a world in which professional standards are disregarded by top-level managers who are preoccupied with maintaining self-promoting images and forming power alliances with other managers. Hard work, commitment to worthwhile and safe products, and even profit making take a back seat to personal survival in the tumultuous world of corporate takeovers and layoffs. It is noteworthy that Jackall's book is based primarily on his study of several large chemical and textile companies during the 1980s, companies notorious for indifference to worker safety (including cotton-dust

---

[16] Robert Jackall, *Moral Mazes: The World of Corporate Managers* (New York: Oxford University Press, 1988), pp. 6, 109 (italics removed).

"I THINK WE CAN PULL OFF A TAKEOVER OF THIS COMPANY— BY THE WAY, WHAT KIND OF PRODUCT DOES IT MAKE?"

TRADER, RAIDER, GREENMAIL AND FASTBUCK = INVESTORS =

Reprinted, with permission, from *Herblock at Large* (Pantheon, 1987)

poisoning) and environmental degradation (especially chemical pollution).

Fortunately, most corporations do not fit Jackall's profile, although one never knows when a company will be acquired by another one with less interest in the product line and an ethical climate than in raiding the employees' pension plan. Several companies that do place a high priority on concern for worthwhile products and ethical procedures are described in *Companies with a Conscience* by Mary Scott and Howard Rothman. One is Quickie Designs, which manufactures wheelchairs.[17] The company was founded in 1980 by Marilyn Hamilton, a school teacher and athlete who two years earlier was paralyzed in a hang-gliding accident. Her desire to return to an active life was frustrated by the unwieldiness of the heavy wheelchairs then available. At her request, two of her friends designed a highly mobile, lightweight, and versatile wheelchair made from aluminum tubing originally developed in the aerospace industry and now used to make hang gliders. The friends created a company that rapidly expanded to make a variety of *innovatively* engineered products for people with disabilities. The company went on to create and support Winners on Wheels, a not-for-profit organization that sponsors sports events for young people in wheelchairs.

Like most of the companies described in *Companies with a Conscience,* Quickie Designs is both relatively small and exceptionally committed to what has been called "caring capitalism." Larger corporations characterized by more intense competition and profit-making pressures face a much greater challenge in maintaining an ethical climate. But many corporations are finding ways to deal with these pressures.

For example, in 1985, Martin Marietta Corporation (now Lockheed Martin), the large aerospace and defense contractor, began an ethics program emphasizing basic values like honesty and fairness, which later expanded to include responsibilities for the environment and for high product quality.[18] Part of the stimulus for the program was public scrutiny of the defense industry, and indeed Martin Marietta was being investigated at that time for improper billings to the government for travel expenses. Lockheed also had a troubled past. In 1972 and 1973, it had engaged in paying bribes to facilitate the sale of its TriStar jumbo jets to Japan Airlines (a story we will revisit as a case study in Chapter 6). Earlier, well-placed connections had also enabled the company to license, manufacture, and sell components of its F-104 Starfighter to airplane builders in Germany (where eventually 175 crashed, killing 85 pilots) and in Japan (where 54 jets were lost).

Nevertheless, the ethics program developed was not merely a reaction designed to avoid legal penalties, but also a concerted effort to institutionalize ethical commitments throughout the newly merged corporation. Specifically, the ethics team drafted a

---

[17] Mary Scott and Howard Rothman, *Companies with a Conscience* (New York: Carol Publishing Group, 1992), pp. 103–17.

[18] Lynn Sharp Paine, "Managing for Organizational Integrity," *Harvard Business Review,* March–April 1994, pp. 106–17.

code of conduct, conducted ethics workshops for managers, and created effective procedures for employees to express their ethical concerns. An ethics network links a central ethics office with ethics representatives appointed at each major facility. In 1991, when the company had about 60,000 employees, some 9,000 confidential employee inquiries or complaints entered the network, and during the following year, 684 investigations were conducted. In the meantime, the company has also produced an ethics training game featuring Dilbert, a popular comic-strip character.

Numerous other examples could be cited, but one will have to do. Texas Instruments (TI) has had a long-standing emphasis on trust and integrity, but during the 1980s, it greatly intensified its efforts to make ethics central to the corporation.[19] In 1987, TI appointed a full-time ethics director, Carl Skooglund, who was then a vice president for the corporation. Skooglund reported to an ethics committee that in turn reported directly to the board of directors. His activities included raising employees' ethical awareness through discussion groups and workshops on ethics, addressing specific cases and concerns in weekly newsletters, and making himself directly available to all employees through a confidential phone line. In this way, Skooglund served as an ethics *ombudsperson,* beyond his role as ethics educator and director of corporate policies on ethics. In all aspects of the program, the emphasis was on supporting ethical conduct, rather than punishing wrongdoers, although it was made clear that unprofessional conduct would not be tolerated.

Not all attempts to establish corporate ethics programs are successful. During the late 1980s, a large defense contractor established a program that included top-level ethics planning, appointment of division-level ethics directors, establishment of new channels for handling complaints, and training programs for employees.[20] Higher management viewed the program as a success because the company avoided scandals faced by competitors, but a group of professional employees assessed the program as a sham intended for public relations and window dressing. The primary difficulty seemed to be a gap between the intentions of top management and the unchanged conduct of senior line managers, a gap that engendered employee cynicism. The company also emphasized a negative approach by requiring employees to sign cards stating they understood the new requirements and by widely publicizing sanctions for specific violations.

---

[19] Francis J. Aguilar, *Managing Corporate Ethics* (New York: Oxford University Press, 1994), pp. 120–35, 140–43.

[20] Ibid., pp. 136–42.

How did this company appear to its clients and the government? Probably quite acceptable; after all, there was an ethics compliance program that made sure all relevant laws were made known. Indeed, the U.S. government would treat transgressions more leniently because of the compliance programs.

What are the defining features of an ethical corporate climate? The preceding examples suggest at least four features. First, ethical values in their full complexity are widely acknowledged and appreciated by managers and employees alike. Responsibilities to all constituencies of the corporation are affirmed—not only to stockholders, but also to customers, employees, and all other stakeholders in the corporation. That does not mean that profits are neglected, nor does it neglect the special obligations that employees of corporations have to promote the interests of the corporation. For the most part, the public good is promoted through serving the interests of the corporation. Nevertheless, the moral limits on profit seeking go beyond simply obeying the law and avoiding fraud. Precisely what those limits are is a matter of ongoing discussion in democracies, for example, concerning such things as tobacco, weapons, and dangerous recreational vehicles.

Second, and implied by the first feature, the use of ethical language is honestly applied and recognized as a legitimate part of corporate dialogue. One way to emphasize this legitimacy is to make prominent a corporate code of ethics. Another way is to explicitly include a statement of ethical responsibilities in job descriptions of all layers of management.

Third, top management must set a moral tone, in words, in policies, and by personal example. Official pronouncements asserting the importance of professional conduct in all areas of the corporation must be backed by support for professionals who work according to the guidelines outlined in professional codes of ethics. Whether or not there are periodic workshops on ethics or formal brochures on social responsibility distributed to all employees, what is most important is fostering confidence that management is serious about ethics. Sometimes the real test arises in connection with mergers and acquisitions: Some ethics programs quietly vanish, others are strengthened, and some appear where none had existed before.

Fourth, there need to be procedures for conflict resolution. One avenue, exemplified by Texas Instruments and Martin Marietta, is to create ombudspersons or designated executives with whom employees can have confidential discussions about moral concerns. Equally important is educating managers on conflict resolution, a topic to which we return in Chapter 5.

In building an ethical corporate climate, it is particularly important not to fall into the trap of relying solely on conveniently *legalistic* compliance strategies. These appear to be much favored by lawyers and executives who can then lay sole blame for organizational failures on individuals who have supposedly acted contrary to the organization's rules, whether these actually promote ethical behavior or not. Specific compliance rules are suitable only in very structured settings, such as purchasing and contracting. What needs to be encouraged is autonomous ethical decision making, as will be explained in Chapter 2.

Social Responsibility Movement

Despite the increasing economic pressures on corporations, the time has never been better to establish an ethical climate. Since the 1960s, a "social responsibility movement" has raised sensitivity within companies concerning product quality and the well-being of workers, the wider community, and the environment. It would be difficult to find a CEO today who would publicly say that all to which his or her company was devoted is profits, that greed is good, and that the environment, workers, and even customers be damned when they get in the way of profits. Such a stance would be self-defeating, for it would alienate support from many constituencies needed for a corporation to prosper. Corporations depend upon and have responsibilities to employees, customers, dealers, suppliers, local communities, and the general public.

Thus, beyond being concerned with employee relations and other internal organizational matters, responsible corporations also strive to be good neighbors by supporting local schools, cultural activities, civic groups, and charities. But frequently, the wider question of how a corporation's product is ultimately used, and by whom, is conveniently put aside because the effects often do not appear nearby or early on, and important questions are therefore not raised. For instance, what happens to used dry-cell batteries? In the United States, nearly three billion of them, along with their noxious ingredients, end up in the municipal waste stream annually. Worldwide, 15 billion are produced per year. What can be done? (A discussion topic at the end of this chapter invites further exploration of this issue.)

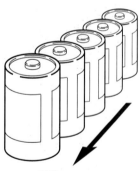

Billions a year

While many corporations are genuinely concerned about what happens to a product once it leaves the factory, others have ready excuses: "We cannot control who buys the product, how it is used, how it is discarded!" Obviously, the task is not easy and usually requires industrywide and government efforts, but socially responsible corporations participate in finding solutions, a task that satisfies even shareholders when common action throughout a particular industry is in the offing, or when the corporation can shine as an industry pioneer. A good example of a corporation's efforts to be in touch with its customers is Cardiac Pacemakers Inc. of St. Paul, Minnesota. Heart patients are invited to the plant so they may share and perhaps alleviate their concerns while employees working on the pacemakers develop a heightened awareness of their responsibilities to turn out high-quality product.

To be sure, the social responsibility movement in business is not without its critics who contend that corporations should concentrate solely on maximizing profits for stockholders and that there are no additional responsibilities to society, customers, and employees. In particular, Nobel Laureate Milton Friedman attacked the social responsibility movement in a widely read article published in 1970, "The Social Responsibility of Business Is to Increase Its Profits." Friedman argued that the paramount, indeed the sole, responsibility of management is to satisfy the desires of stockholders who entrust corporations with their money in order to maximize return on their investment. Management acts irresponsibly and violates stockholders' trust when it adopts further social goals, such as protecting the environment, training disadvantaged workers, using affirmative action hiring practices, or making philanthropic donations to local communities or the arts. The responsibility of managers is "to conduct the business in accordance with their [stockholders'] desires, which generally will be to make as much money as possible while conforming to the basic rules of the society, both those embodied in law and those embodied in ethical custom."[21]

Our customers are not the people buying cars, but the people buying our stock.
—Attributed to Lee Iacocca

Ironically, Friedman's allusion to heeding "ethical custom" would seem to permit or even invite recognition of many of the wider corporate responsibilities he inveighs against. For instance, suppose that in our society the public expects corporations to contribute to wider community goods and to protect the environment, and that becomes a moral custom (as indeed it largely has). It seems clear, however, that by "ethical custom,"

---

[21] Milton Friedman, "The Social Responsibility of Business Is to Increase Its Profits," *The New York Times Magazine,* September 13, 1970.

Friedman means only refraining from fraud, deception, and corruption.[22]

Friedman's view is ultimately self-defeating. As quickly as the public learns that corporations are indifferent to anything but profit, it will tend to respond by passing increasingly restrictive laws that make profit-making difficult. Conversely, when the public perceives corporations as having wider social commitments, it is more willing to cooperate with them to assure reasonable regulations, and to selectively purchase products from such socially responsible corporations. Even many investors will be more likely to stay with companies whose ethical commitments promise long-lasting success in business.

Sound ethics and good business go together, for the most part and in the long run. Hence, at a fundamental level, the moral roles of engineers and managers are complementary and symbiotic, rather than opposed. As managers, engineers remain professionals whose primary moral responsibility is to provide safe and useful products that are profitable. Admittedly, the transition to management does involve adjustments and shifts in emphasis. Moreover, higher management tends to be dominated by a culture that sometimes clashes with the culture of professional work of engineers.[23] As a result of their different experience, education, and roles, higher management tends to emphasize corporate efficiency and productivity—the bottom line. Engineers and other professionals tend to emphasize excellence in creating useful, safe, and quality products. But these differences should be a matter of emphasis rather than opposition.

In order to make their own voices heard more clearly, engineers and scientists have formed societies to discuss and promote social responsibility in their professions. Examples are the Federation of American Scientists (founded after World War II by participants in the Manhattan Project), the Computer Professionals for Social Responsibility, and the IEEE Society on Social Implications of Technology.

Senses of "Moral Responsibility"

We have been speaking of the responsibilities of corporations and the engineers they employ, but *responsibility* is an ambiguous

---

[22] Milton Friedman, *Capitalism and Freedom* (Chicago: University of Chicago Press, 1963), p. 133. See Thomas Carson, "Friedman's Theory of Corporate Social Responsibility," *Business and Professional Ethics Journal* 12 (Spring 1993), pp. 3–32. For critiques of Friedman's view, see Peter Drucker, *An Introductory View of Management* (New York: Harper & Row, 1974), p. 61; Robert C. Solomon, *Ethics and Excellence: Cooperation and Integrity in Business* (New York: Oxford University Press, 1992), p. 120.

[23] Joseph A. Raelin, *The Clash of Cultures: Managers Managing Professionals* (Boston: Harvard Business School Press, 1991).

term. In distinguishing each sense, we will begin with how the term *responsibility* is applied to individuals and then how a similar sense can be applied to corporations.

Moral Responsibility

Moral judgments are involved whenever moral responsibility is ascribed to individuals or corporations, but the judgments may be of various types.[24] They might ascribe (1) a virtue, (2) obligations, (3) general moral capacities, (4) liabilities and accountability for actions, or (5) blameworthiness or praiseworthiness.

1. When we say someone is a responsible person, we mean to ascribe a general moral *virtue* to the person, an admirable feature of character. We mean that he or she is regularly concerned to do the right thing, is conscientious and diligent in meeting obligations, and can be counted on to carry out duties or be considerate of others. This is the sense in which professional responsibility is the central virtue (morally desirable feature) of engineers. In this same sense, we can speak of some corporations as manifesting the virtue of responsibility—of moral or ethical commitment to meet its duties.

2. We speak of persons as having moral responsibilities. In this sense, responsibilities are simply *obligations* or duties. Some of those responsibilities are shared by us all: for example, the responsibilities to be truthful, to be fair, and to promote justice. Others relate only to people performing within certain social roles or professions. For example, a safety engineer might have responsibilities for making regular inspections at a building site, or an operations engineer might have special responsibilities for identifying potential benefits and risks of one system as compared to another. The virtue of professional responsibility implies the conscientious effort to meet the responsibilities inherent in one's work. In this same sense, we ascribe responsibilities to corporations. Their responsibilities are simply the general obligations they have.

3. Sometimes when we ascribe responsibility to a person viewed as a whole rather than in respect to a specific area of his or her conduct, we have in mind an active *capacity* for knowing how to act in morally appropriate ways. In this sense, young children are not yet morally responsible. They gradually become so as they mature and learn how to be responsive to the needs and interests of others. Obviously, the virtue of professional responsibility implies responsibility in this capacity sense. Corporations, like

---

[24] H. L. A. Hart, *Punishment and Responsibility* (Oxford, England: Clarendon, 1973), pp. 211–30.

individuals, have the capacity for morally responsible agency because it is intelligible to speak of the corporation as acting. The actions of the corporation, of course, are performed by individuals and subgroups within the corporation, according to how the flow chart and policy manual specify areas of authority.[25]

4. In a closely related sense, *responsible* often means *accountable,* or answerable for meeting particular obligations. In this sense, to say individuals are responsible for actions means they can be "held to account" for them: that is, they can be called upon to explain why they acted as they did, to provide excuses or justification if appropriate, and to be open to commendation or censure, praise or blame, or demands for compensation. We also hold ourselves accountable for our own actions, responding to them with emotions of self-esteem or shame, self-respect or guilt. Corporations are accountable in this same sense—accountable to the general public, to their employees and customers, and to their stockholders.

   Of course, sometimes there are legitimate excuses for failures to meet responsibilities. Roughly, we are responsible for *voluntary actions,* that is, actions that (1) we knew or should have known were right or wrong, and (2) were not done under coercion or uncontrollable compulsions.

5. In contexts where it is clear that accountability for wrongdoing is at issue, *responsible* becomes a synonym for *blameworthy.* In contexts where it is clear that right conduct is at issue, *responsible* is a synonym for *praiseworthy.* Thus, the question "Who is responsible for designing the antenna tower?" may be used to ask who is blameworthy for its collapse or who deserves credit for its success in withstanding a severe storm. Corporations, too, are blamed or praised according to how well they meet their responsibilities.

Causal and Legal Responsibility

There are two other concepts of responsibility that should not be confused with moral responsibility in any of its five preceding senses. First, *causal responsibility* consists simply in being a cause of some event. In this sense, we speak of lightning as being responsible for starting a fire. People can be causally responsible for an event without necessarily being morally responsible for it. For example, a two-year-old child may cause a fire while playing with matches, but it is the parents who left the matches within the child's reach who are morally responsible for the fire.

---

[25] Peter A. French, *Corporate Ethics* (New York: Harcourt Brace, 1995).

Second, *legal responsibility* should also be distinguished from moral responsibility. An engineer or engineering firm can be held legally responsible for harm that was so unlikely and unforeseeable that little or no moral responsibility is involved.

One famous court case involved a farmer who lost an eye when a metal chip flew off the hammer he was using.[26] He had used the hammer without problems for 11 months before the accident. It was constructed from metals satisfying all the relevant safety regulations, and no specific defect was found in it. The manufacturer was held legally responsible and required to pay damages. The basis for the ruling was the doctrine of *strict legal liability,* which does not require proof of defect or negligence in design. Yet surely the manufacturer was not morally guilty or blameworthy for the harm done. It is morally responsible only insofar as it has an obligation (based on the special relationship between it and the farmer created by the accident) to help repair the harm caused by the defective hammer.

Discussion Topics

1. Milton Friedman argues that the sole responsibility of managers is to stockholders, to maximize their profits within the bounds of law and without committing fraud. An alternative view is that managers have responsibilities to all individuals and organizations that make contracts with a corporation or otherwise are directly affected by them. These stakeholders include employees, customers, suppliers, government regulators, and members of the local communities in which the corporation locates its facilities.[27] Clarify what you see as the implications of these alternative views as they apply to decisions about relocating a manufacturing facility in order to lower costs for workers' salaries. Then, present and defend your view as to which of these positions is the more defensible morally.

2. Owners of a corporation enjoy the advantages of limited liability for corporate debt or obligations, but a corporation is more than an inanimate object. When corporations were first created as legal entities, lawyers drawing on Roman law endowed these creations with personal attributes. Today, some would hold individual persons (employees, managers) responsible for the actions of the corporation, while others would find it appropriate to hold the corporation as an entity responsible. Both positions

---

[26] Richard C. Vaughn, *Legal Aspects of Engineering,* 3rd ed. (Dubuque, IA: Kendall/Hunt, 1977), pp. 41–47.

[27] James J. Brummer, *Corporate Responsibility and Legitimacy* (New York: Greenwood Press, 1991), pp. 144–164; Ronald M. Green, *The Ethical Manager* (New York: Macmillan, 1994), pp. 25–42.

apply in a legal sense. Thus, an individual can be declared guilty (and subject to criminal prosecution) and a corporation can be held liable (and be subject to monetary penalties) for the same harmful action.[28] How would you assess proposals that it should be possible to sentence a corporation to "death" (having its charter revoked) for particularly flagrant behavior?[29]

3. Three engineers test a product together but fail to disclose a defect that will not be noticed until much later. Can blame for this omission be apportioned among the engineers according to some scale, or is there no difference in degree of complicity since they all kept silent?[30]

References

General Sources for Engineering Ethics

From the extensive literature on applied ethics, only the following journals, books, and web sites are listed here because of their comprehensive coverage of engineering ethics.

Journals

1. *Science and Engineering Ethics*
2. *Business & Professional Ethics Journal*
3. *Professional Ethics*

Books

1. Davis, Michael. *Thinking Like an Engineer.* New York: Oxford, 1998.

2. Fleddermann, Charles B. *Engineering Ethics.* Upper Saddle River, NJ: Prentice Hall, 1999.

3. Harris, Charles E., Jr.; Michael S. Pritchard; and Michael J. Rabins. *Engineering Ethics: Concepts and Cases.* Belmont, CA: Wadsworth, 1995.

4. Johnson, Deborah G. *Ethical Issues in Engineering.* Englewood Cliffs, NJ: Prentice Hall, 1991.

5. Schaub, James H., and Karl Pavlovic, eds. *Engineering Professionalism and Ethics.* New York: John Wiley, 1983.

6. Schlossberger, Eugene. *The Ethical Engineer.* Philadelphia: Temple University Press, 1993.

---

[28] Roland Schinzinger and Mike W. Martin, "Shared Responsibility for New Technologies: Engineers and Their Corporations," in *A Delicate Balance: Technics, Culture, and Consequences,* Proceedings of an IEEE-SSIT Conference, Los Angeles, CA, 1989.

[29] Russell Mokhiber, "Death Penalty for Corporations Comes of Age," *Business Ethics,* November/December 1998, pp. 7–8.

[30] Roland Schinzinger and Mike W. Margin, "On Complicity: The Ethical Dimensions of Co-Responsibility in Technology," in *Engineering Education 2000,* IEEE-IGIP-ASSEE Symposium Proceedings, Vienna: Leuchtturm Verlag, 1990.

7. Unger, Stephen H. *Controlling Technology.* 2nd ed. New York: John Wiley, 1994.

8. Vesilind, P. Aarne; Alastair Gunn. *Engineering, Ethics, and the Environment.* New York: Cambridge University Press, 1998.

9. Whitbeck, Caroline. *Ethics in Engineering Practice and Research.* New York: Cambridge University Press, 1998.

Web Sites

Here is a sampling of relevant web sites. Most allow direct transfer to each other and to many good sites not listed below.

1. National Institute for Engineering Ethics, http://www.niee.org

2. National Society of Professional Engineers, http://www.nspe.org

3. The Online Ethics Center for Engineering and Science http://www.onlineethics.org

2

Moral Reasoning and Ethical Theories

An ethical theory is an attempt to provide a perspective on moral responsibilities, and morality in general, that expresses our carefully considered moral convictions, identifies what is morally fundamental, is clear and comprehensive, and provides a framework for resolving moral dilemmas. During the past three centuries, three ethical theories have been especially influential: utilitarianism, rights ethics, and duty ethics. *Utilitarianism* holds that we ought to maximize the overall good, taking into equal account all those affected by our actions. *Rights ethics* says we ought to respect human rights. *Duty ethics* says we have duties to respect persons' autonomy (self-determination).

No one of these theories has won a consensus, and indeed each theory has different versions. Nevertheless, each theory captures important insights that make them worthy of attention. After explaining their key ideas, we will explore alternative responses to them and in doing so introduce three further views that are currently receiving considerable attention: John Rawls's theory of justice, virtue ethics, and pragmatism. Following this discussion, we relate ethical theories to self-interest, respect for customs, and religious beliefs. There we also indicate how moral motivation—motivation in terms of rights, duties, virtues, and so on—plays an important role in professionalism.

Truthfulness

In comparing and contrasting the ethical theories, it will be helpful to focus on truthfulness as an illustration. Not only is truthfulness a basic moral consideration of which any adequate ethical theory would have to take account, but it pertains to all aspects of engineering. Not surprising, then, truthfulness enters frequently into the cases discussed by the National Society of Professional Engineers (NSPE). Here are summaries of cases

39

from a recent edition of the NSPE *Opinions of the Board of Ethical Review.*[1]

1. A city hires an engineer to design a bridge, and the engineer in turn subcontracts some key design work to a second engineer. Months after the bridge is completed, the first engineer submits the design to a national design competition where it wins an award, but he fails to credit the work of the second engineer. (Case No. 92-1)

2. An engineer who is an expert in hydrology and a key associate with a medium-sized engineering consulting firm gives the firm her two-week notice, intending to change jobs. The senior engineer-manager at the consulting firm continues to distribute the firm's brochure, which lists the woman as an employee of the firm. (Case No. 90-4)

3. A city advertises a position for a city engineer/public works director, seeking to fill the position before the incumbent director retires in order to facilitate a smooth transition. The top candidate is selected after an extensive screening process, and on March 10 the engineer agrees to start April 10. By March 15, the engineer begins to express doubts about being able to start on April 10, and after negotiations, the deadline is extended to April 24, based on the firm commitment by the engineer to start on that date. On April 23, the engineer says he has decided not to take the position. (Case No. 89-2)

4. An engineer working in an environmental engineering firm directs a field technician to sample the contents of storage drums on the premises of a client. The technician reports back that the drums most likely contain hazardous waste, and hence require removal according to state and federal regulations. Hoping to advance future business relationships with the client, the engineer merely tells the client the drums contain "questionable material" and recommends their removal, thereby giving the client greater leeway to dispose of the material inexpensively. (Case No. 92-6)

NSPE discusses these cases by citing the following "Rule of Practice" stated in the NSPE Code of Ethics: "Engineers shall be objective and truthful in professional reports, statements or testimony. They shall include all relevant and pertinent information in such reports, statements or testimony."[2] This *truthfulness*

---

[1] National Society of Professional Engineers, *Opinions of the Board of Ethical Review,* Vol. 7 (Alexandria, VA: NSPE, 1994); also see the web site at www.nspe.org/eh-home.asp.

[2] Ibid.

rule is affirmed by all the major ethical theories, but for different reasons linked to the "moral bottom line" of the theory: utility, rights, or duties.

Utilitarianism

Utilitarians say that we should understand truthfulness, like all other moral considerations, in terms of its contribution to good consequences. Utilitarianism is the view that we ought to produce the most good for the most people, giving equal consideration to everyone affected. The standard of right conduct is maximization of good consequences. "Utility" is sometimes used to refer to these consequences, and other times it is used to refer to the balance of good over bad consequences.

At first glance, this seems simple enough. But what is the goodness that is to be maximized? And how is the "production" of goodness to be assessed: with respect to each action or with respect to the consequences of general rules about actions? Depending on how these questions are answered, utilitarianism can be developed in different directions.

Rule-utilitarianism focuses on the overall consequences of rules, such as the truthfulness rule, rather than of individual actions. We ought always to act on those rules that if generally adopted would produce the most good for the most people. According to Richard Brandt (1910–1997), rules should be considered in sets that he calls *moral codes.*[3] A moral code is justified when it is the *optimal code* that, if adopted and followed, would maximize the public good more than alternative codes would. The codes may be societywide standards or special codes for a profession like engineering. Individual actions are right when they conform to such rules.

Act-utilitarianism, by contrast, focuses on each action and regards rules as at most useful rules of thumb. An act is right if it is likely to produce the most good for the most people involved in the particular situation. The truthfulness rule, as well as everyday maxims such as "Tell the truth" and "Do not deceive," are only rough guidelines. According to John Stuart Mill (1806–1873), these maxims are useful rules of thumb that summarize past human experience about the types of actions that usually maximize utility.[4] But the rules should be broken whenever doing so will produce the most good in a specific situation.

There are debates over precisely how much rule- and act-utilitarianism differ from each other, but sometimes they seem

---

[3] Richard B. Brandt, *A Theory of the Good and the Right* (Oxford: Clarendon Press, 1979).

[4] John Stuart Mill, *Utilitarianism, with Critical Essays,* ed. Samuel Gorovitz (Indianapolis, IN: Bobbs-Merrill, 1971).

clearly to point in different directions. Certainly, utilitarians see the difference as crucial. Rule-utilitarians accuse act-utilitarians of opening too many "loopholes" licensing exceptions, in particular deceiving when doing so promotes more good than bad. Act-utilitarians reply that this is precisely the true spirit of utilitiarianism and that fixation on rules leads to moral rigidity rather than maximum good.

> It is quite compatible with the principle of utility to recognize the fact that some kinds of pleasure are more desirable and more valuable than others.
> —John Stuart Mill, *Utilitarianism, with Critical Essays*

Theories of Good

If the standard of right action is maximizing goodness, whether measured in terms of rules or separate actions, what is goodness? Mill believed that happiness is the only *intrinsic good,* that is, something good in and of itself, or desirable for its own sake. All other good things are *instrumental goods* in that they provide means ("instruments") for gaining happiness.

What is happiness? Consistent with Mill, we might think of it as a combination of (*a*) a life rich in deeply satisfying pleasures, mixed with some inevitable pains, and (*b*) a pattern of activities and relationships that one can affirm as valuable overall, as the way one wants one's life to be. Exactly what this amounts to is a highly personal matter, and one that each of us frequently has difficulty in discerning. In Mill's view, a happy life is composed of many pleasures in great variety, mixed with some inevitable brief pains. The happiest life is also rich in *higher pleasures.* Higher pleasures are preferable in quality or *in kind* to other pleasures. For example, Mill contended that the pleasures derived through love, friendship, intellectual inquiry, creative accomplishment, and appreciation of beauty are inherently better than the bodily pleasures derived from eating, sex, and exercise. More recent utilitarians have included the virtues on their list of intrinsic goods, including the virtue of truthfulness or honesty.

By contrast, Richard Brandt argues that things such as love and creativity are good because they satisfy rational desires. *Rational desires* are those we would have and approve of if we scrutinized our desires in light of all relevant information about the world and our own psychology. Some self-destructive desires, such as the desire to use dangerous drugs, are not rational since, if we saw their full implications, we would not approve of them.

Mill and Brandt share in common the attempt to impose some objective standards on what counts as good. Other utilitarians, especially economists, are concerned with difficulties about how to identify and measure desires and the pleasures they yield. They seek a more public but also more subjective way to determine the good. Economists base their cost–benefit analyses on the preferences that people express through their buying habits. In this version, utilitarianism becomes the view that right actions produce the greatest satisfaction of the preferences of people affected. The difference between objective and subjective standards of good is as important as the difference between rule- and act-utilitarianism in thinking about when and why truth-telling is obligatory.

Rights Ethics

According to rights ethics, human rights—not good consequences—are fundamental. Human rights can be viewed as a fundamental moral authority that all human beings possess so they may live their lives and exercise their liberty. Acts of respect for human rights are obligatory, regardless of whether they always maximize good. Rights ethicists tend to understand truthfulness in terms of its contribution to liberty, especially within relationships based on trust, simply because truth itself is essential to being able to exercise liberty in order to pursue our many goals.

Rights ethics, like utilitarianism, seems at first like a simple theory, but it takes on greater complexity as we inquire into different types of rights and how rights are to be balanced against each other when they conflict.

Liberty Rights

John Locke (1632–1704) argued that to be a person entails having human rights to life, liberty, and the property generated by one's labor. His views had an enormous impact at the time of the French and American revolutions and provide the moral foundation of contemporary American society. The words in the *Declaration of Independence* are not far from his own: "We hold these truths to be self-evident, that all men are created equal, that they are endowed by their Creator with Unalienable Rights, that among these are Life, Liberty, and the pursuit of Happiness." In writing these words, Thomas Jefferson primarily had in mind male white landowners, but his words were prescient in founding the U.S. political-legal system that would eventually embrace all people as having human rights.

Locke's own version of a human rights ethics was highly individualistic. He viewed rights primarily as entitlements that prevent other people from meddling in one's life. These are referred to as *liberty rights* (or "negative rights") that place duties on other people not to interfere with one's life. He thought of property as whatever we gain by "mixing our labor" with things, by contrast with modern legal definitions of property as what the law and government specify as to how we can gain and use material things. The individualistic aspect of Locke's thought is reflected in the contemporary political scene in the libertarian outlook, with its emphasis on protection of private property and the condemnation of welfare systems. Libertarians take a harsh view of taxes and government involvement beyond the bare minimum necessary for national defense and preservation of free enterprise. This perspective contrasts with a second version of human rights ethics.

Liberty and Welfare Rights

This second version of rights ethics conceives of human rights as intimately related to communities of people. A. I. Melden (1910–1991) argued that having moral rights presupposes the capacity to show concern for others and to be accountable within a moral community.[5] The extent of rights, in his view, always has to be determined in terms of interrelationships among persons. Melden's account allows for more "positive" *welfare rights* to community benefits needed for living a minimally decent human life (when one cannot earn those benefits on one's own and when the community has them available). Thus, it lays the groundwork for recognizing a social welfare system such as the United States currently has.

Moral rights encompass more than human rights. Some arise from special relationships and roles that people might have. Thus, engineers and other professionals are required to be truthful in performing their work because of the rights of their employers and clients created by contracts. Again, many of the special rights we will examine later arise within institutions and professions, such as the right of engineers to warn the public about unsafe technological products. But rights ethicists seek to justify special rights by reference to human rights. Thus, according to Melden, contracts and other types of promises create special rights because people have human rights to liberty that are violated when the understandings and commitments specified in contracts and promises are violated.

Duty Ethics

Some philosophers think of duty ethics as the mirror image of rights ethics. For most rights, there are corresponding duties.

---

[5] A. I. Melden, *Rights and Persons* (Berkeley, CA: University of California Press, 1977).

Thus, if you have a right not to be deceived, then I have a duty not to deceive you. Again, if you have a right to life, then I have a duty not to kill you by marketing deadly products; and if I have a duty to respect your freedom, then you have a right not to be interfered with. Nevertheless, duty ethics takes duties rather than rights as fundamental, and in doing so it shifts the emphasis to what we owe morally to others.

Respect for Autonomy

Immanuel Kant (1724–1804) is the most famous of the ethicists who regard duties as fundamental.[6] In his view, right actions are those required by a list of duties, such as be truthful, be fair, make reparation for harm done, and show gratitude for kindness extended by others. There are also duties to ourselves, such as seek to improve one's own character and talents and not to commit suicide. More recent duty ethicists attempt to identify a comprehensive list of basic duties. For example, Bernard Gert offers this list of 10 fundamental duties: "1. Don't kill. 2. Don't cause pain. 3. Don't disable. 4. Don't deprive of freedom. 5. Don't deprive of pleasure. 6. Don't deceive. 7. Keep your promise. 8. Don't cheat. 9. Obey the law. 10. Do your duty [referring to work, family, and other special responsibilities]."[7]

> Moral self-knowledge, which requires one to penetrate into the unfathomable depths and abyss of one's heart, is the beginning of all human wisdom.
>
> —Immanuel Kant,
> *Foundations of the Metaphysics of Morals*

What unifies such lists of duties, and how do we determine what our duties are? According to Kant, each duty expresses respect for persons. People deserve respect precisely because they have inherent worth as rational beings who have the capacity for (moral) autonomy. *Autonomy* means governing our own lives in light of universal moral principles—principles that apply equally to all people, rules that as rational agents we freely accept as binding on us. Hence, respect for persons amounts to respect for their autonomy and for their efforts to meet their duties. To deceive others, for example, is to undermine their ability to carry out their plans based on available truths and within

---

[6] Immanuel Kant, *Foundations of the Metaphysics of Morals, with Critical Essays,* ed. Robert Paul Wolff (Indianapolis, IN: Bobbs-Merrill, 1969).

[7] Bernard Gert, *Morality* (New York: Oxford University Press, 1988), p. 157.

relationships based on trust. Equally important is respect for oneself, that is, conscientiously seeking to fulfill our duties to ourselves.

Deceivers, Kant suggests, use other people as "mere means" to their own selfish purposes, rather than respecting them as rational beings with desires and needs. They are also betraying their own integrity and failing to respect themselves as morally concerned persons. Violent acts such as murder, rape, and torture are even more flagrant ways of treating people as mere objects serving our own purposes. Of course, we need to "use" one another all the time: Business partners, employers and employees, and faculty and students use each other to obtain their personal ends. Immorality occurs when we "merely use" others, when we reduce them to being *mere* means to our ends, and nothing more.

Prima Facie Duties

A major difficulty with Kant's view is that he confused the ideas of (1) rules being universally applicable to all rational agents with (2) rules being exceptionless. He thought principles of duty were *absolute* in the sense of never having justifiable exceptions. In doing so, he failed to be sensitive to how principles of duty can conflict with each other, thereby creating moral dilemmas. Contemporary duty ethicists recognize that many moral dilemmas are resolvable only by recognizing some valid exceptions to simple principles of duty. Thus, "Do not deceive" is a duty, but it has exceptions when it conflicts with the moral principle "Protect innocent life." One ought to deceive a kidnapper if that is the only way to keep a hostage alive until the police can intervene. Principles of duty that can have justified exceptions are called *prima facie duties,* a widely used expression introduced by David Ross.[8] Most duties are in fact prima facie ones, and this expression also applies to rights and other moral principles to indicate that they too have limits.

How do we tell which duties should override others when they come into conflict? Ross emphasized the importance of careful reflection on each situation, weighing all relevant duties in light of all the facts, and trying to arrive at a sound judgment or intuition. He also stressed that some principles, such as "Do not kill" and "Protect innocent life," clearly involve more pressing kinds of respect for persons than other principles, such as "Don't lie." By contrast, most duty ethicists, like Bernard Gert, attempt to minimize the need for (often variable) intuitions by engaging in pub-

---

[8] W. D. Ross, *The Right and the Good* (Oxford: Oxford University Press, 1946).

lic reasoning aimed at discovering a consensus among rational agents about justified exceptions. Of course, very often in engineering, the demand for protecting safety requires being truthful, and in general, moral rules are mutually reinforcing the more they come into conflict with each other.

Discussion Topics

1. Apply utilitarianism, duty ethics, and rights ethics in resolving the following moral problems. Be sure to consider alternative versions of each theory, such as act-utilitarianism and rule-utilitarianism. Do the theories lead to the same or different answers to the problems?

    (a) George had a bad reaction to an illegal drug he accepted from friends at a party. He calls in sick the day after, and when he returns to work the following day, he looks ill. His supervisor asks him why he is not feeling well. Is it morally permissible for George to lie by telling his supervisor that he had a bad reaction to some medicine his doctor prescribed for him?

    (b) Jillian was aware of a recent company memo reminding employees that office supplies were for use at work only. Yet she knew that most of the other engineers in her division thought nothing about occasionally taking home notepads, pens, computer disks, and other office "incidentals." Her eight-year-old daughter had asked her for a company-inscribed ledger like the one she saw her carrying. The ledger costs less than $20, and Jillian recalls that she has probably used that much from her personal stationery supplies during the past year for work purposes. Is it all right for her to take home a ledger for her daughter without asking her supervisor for permission?

    (c) Robert is a third-year engineering student who has been placed on probation for a low grade point average, even though he knows he is doing the best work he can. A friend offers to help him by sitting next to him and "sharing" his answers during the next exam. Robert has never cheated on an exam before, but this time he is desperate. Should he accept his friend's offer?

    (d) Because he had been mugged before, Bernard Goetz (who happens to be an engineer) illegally carried a concealed revolver when he rode the New York subway. When several young men confronted him in a threatening way, asking for money, he drew the revolver and fired several shots that resulted in permanent injuries. Did his right to life and his right to defend himself justify his acts of (1) carrying the revolver and (2) using it as he did?

2. Consult the writings of a major ethicist and summarize the main ideas of the theories involved.[9]

3. It has been said that most (individual) engineers tend to adopt utilitarian positions while ethics review boards of their societies adopt rights-based (deontological) positions.[10] Can you find some NSPE decisions[11] that would verify this observation?

Testing and Refining Ethical Theories

Utilitarianism, rights ethics, and duty ethics all have many defenders, although none of these theories has won a consensus. This lack of consensus about theory does not imply that the theories are unhelpful in thinking about engineering ethics. There are wide areas in which the ethical theories agree, for example, in justifying the truthfulness rule, and each theory can be refined in light of practical cases.

The most obvious response to disagreements over ethical theories is to continue the search for a fully and widely convincing ethical theory by refining further one of the above theories. Because ethical theories attempt to provide comprehensive understanding of ethics, the study of engineering ethics and other branches of practical ethics (medical, business, etc.) can be made part of this ongoing search. Ethical theories are tested and refined using the following tests, which are implied by the very aims in developing ethical theories.

First, the theory must be clear and formulated with concepts that are coherent and applicable. Second, it must be internally consistent in that none of its tenets contradicts any other. Third, neither the theory nor its defense can rely upon false information. Fourth, it must be sufficiently comprehensive to provide guidance in specific situations of interest to us. Fifth, and most important, it must be compatible with our most carefully considered moral convictions (judgments, intuitions) about concrete situations. For example, if an abstract ethical theory said it was all right for engineers to engage in rampant untruthfulness and to create extremely dangerous products without the public's informed consent, that would show the theory is flawed.

In our view, each of the theories—in some of their versions and at least in a substantial degree—might ultimately pass all these tests. Rule-utilitarianism, rights ethics, and duty ethics can be

---

[9] Helpful additional secondary sources are Peter Singer, ed., *A Companion to Ethics* (Oxford: Basil Blackwell, 1991); Lawrence C. Becker, ed., *Encyclopedia of Ethics* (New York: Garland, 1992), second edition forthcoming.

[10] K. L. Carper, "Engineering Code of Ethics: Beneficial Restraint on Consequential Morality," *Journal of Professional Issues in Engineering Education and Practice* 117 (July 1991).

[11] NSPE, *Opinions of the Board of Ethical Review.*

refined so as to capture our most carefully considered convictions. Thus, utilitarians can refine their theory of good and articulate more precise rules; rights ethicists can sharpen their accounts of rights; duty ethicists can refine their list of duties; and all three can attempt to articulate priorities among principles. As authors, we have reservations about whether act-utilitarianism is a sound moral theory because it provides too many "loopholes" sanctioning unfair and unjust conduct, but even that theory admits of considerable refinement according to what theory of good is adopted. We also have doubts about libertarianism as providing too limiting a moral perspective that underestimates the importance of community, but once again, libertarians can refine their view, this time by complementing their theory of rights with an appreciation of the value of voluntary community involvement.

It should come as no surprise that ethical theories will tend to converge in many areas. No sound ethical theory can justify wanton untruthfulness, torture, rape, and exploitation, precisely because the standard for soundness includes ruling such things out as impermissible. Moreover, while there will always be moral disagreements, including some fundamental ones, those differences can be expressed within each of these types of general theory. For example, utilitarians disagree as much among themselves, for example, as they do with duty or rights ethicists. For the purposes of engineering ethics, all of them converge toward similar moral principles, even though they offer

Just as each thing
gives full play
to its natural gifts—

the bamboo those
of the bamboo,
the pine trees those
of the pine trees—

so the good of
the human being
consists in
giving expression
to the innate nature
of its humanity.

Kitaro Nishida
(1870–1945)
from An Inquiry
into the Good

different justifications for those principles. It follows that for practical purposes, it matters little whether we adopt duties or rights as the starting point for moral reflection. Thus, we will draw freely on the language of duties, rights, and utility wherever it aids practical reflections on moral dilemmas in engineering.

John Rawls: Two Principles of Justice

As one noteworthy example of refining a theory, consider the work of John Rawls, whom many believe to be the leading ethicist of our time. In *A Theory of Justice,* Rawls argues for a general perspective in the Kantian duty-ethic tradition. He condenses the fundamental duties applicable to professions and other social institutions down to two principles of justice: (1) each person is entitled to the most extensive amount of political liberty (to vote, engage in free speech, pursue religious convictions, etc.)

> A conception of justice cannot be deduced from self-evident premises or conditions on principles; instead, its justification is a matter of the mutual support of many considerations, of everything fitting together into one coherent view.
>
> —John Rawls, *A Theory of Justice*

compatible with an equal amount for others, and (2) differences in social power and economic benefits are justified only when they are likely to benefit everyone, including members of the most disadvantaged groups.[12] The first principle, which Rawls says has priority and must be satisfied before the second one, is a principle of respect for individuals of the sort that rights ethicists and duty ethicists assert. The second principle captures the importance of good consequences that utilitarians emphasize. Indirectly, the two principles imply truthfulness insofar as it promotes basic liberties and economic justice.

Rawls's way of arguing for his two principles also draws on Kantian ideas of respect for rational autonomy, developed within another tradition called "social contract ethics." Thus, Rawls suggests that valid principles of duty are those upon which all rational persons would voluntarily agree autonomously in an imaginary "contracting" situation. Rational persons in this hypothetical situation have general information (for example, about human psychology and economics) but they are imagined to lack all specific knowledge about themselves—for example, about

---

[12] John Rawls, *A Theory of Justice* (Cambridge, MA: Harvard University Press, 1971), p. 60.

their particular desires, intelligence, achievements, race, and gender. This assures they will not choose principles based on biases in their current situation.

Rawls argues that in the ideally fair contracting situation, rational agents would agree on the two principles to structure our institutions and professions. The first principle is most important and should be satisfied first. Without basic liberties, no other economic or social benefits can be sustained in the long run. The second principle is also very important, however. It insists that allowing some people great wealth and power is justified only when all other groups benefit, especially the most economically disadvantaged groups. Thus, it might be argued that allowing differences of this sort within the free enterprise system is permissible insofar as it provides the capital needed for businesses to prosper, thereby providing job opportunities and taxes to fund a welfare system to help the poor.

Rawls's theory has many defenders but also many critics. Some object that his second principle is too strong and that all that a just society requires is a more minimal welfare system, rather than a constant effort to raise the economic situation of the poor. Others object that rational persons need not be as risk averse as Rawls assumes, and hence might be willing to take greater chances of not succeeding economically and socially. Still other critics contend that his attempt to describe a fair contracting situation is confused because none of us can imagine ourselves without any of our particular features, such as our gender and cultural background.

Once again, then, we see in this debate that ethical theory does not silence moral disagreement. Nevertheless, Rawls's theory illustrates how the attempt to articulate general moral principles can enrich moral debate, including debate about the wider economic settings in which engineering is embedded. Nevertheless, some ethicists, in particular virtue ethicists and pragmatists, are wary of the search for general rules.

Virtues, and Pragmatism

Some philosophers have recently moved away from the rule-oriented approach of the utilitarianism, rights ethics, and duty ethics, toward an ethics of virtue and character. Other philosophers downplay all ethical theories while stressing pragmatic reflection within specific situations. Let us consider each of these approaches.

Virtue Ethics

Whereas rule-utilitarianism, duty ethics, and rights ethics focus on principles of right actions and rules about right conduct, *virtue ethics* focuses first of all on the kinds of persons we should aspire to be. It focuses on character, on patterns of virtues and vices. *Virtues* are desirable ways of relating to other individuals, groups, and organizations. They are desirable habits or tendencies of motives, attitudes, and emotions, as well as right conduct. *Vices* are undesirable habits and tendencies. By extension, we can also speak of the character of organizations, that is, the patterns of virtues and vices that are manifested by management, employees, and corporate policies.

Aristotle: Virtue and the Golden Mean

In his *Nicomachean Ethics,* Aristotle (384–322 B.C.) defined the moral virtues as tendencies, acquired through habit formation, to reach a proper balance between extremes in conduct, emotion, desire, and attitude. To use the phrase inspired by his theory, virtues are tendencies to find *the Golden Mean* between the extremes of too much (excess) and too little (deficiency) with regard to particular aspects of our lives. Thus, truthfulness is the mean between revealing all information in violation of tact and

> The epithets *sociable, good-natured, humane, merciful, grateful, friendly, generous, beneficent,* or their equivalents, are known in all languages, and universally express the highest merit, which *human nature* is capable of attaining.
> —David Hume, *An Enquiry Concerning the Principles of Morals*
>
> Courage is the most important virtue. Without it you cannot practice any other virtue consistently.
> —Maya D'Angelou on Courage

confidentiality (excess) and being secretive or lacking in candor (deficiency) in dealing with truth. Again, courage is the appropriate middle ground between foolhardiness (the excess of rashness) and cowardice (the deficiency of self-control) in confronting dangers. The most important virtue is practical wisdom—morally good judgment—which enables one to discern the mean for all the other virtues.

Virtues enable us to pursue a variety of social goods within a *community*—a concept that was especially important for citizens of ancient Greek city-states, since the city-state's survival depended on close cooperation of its citizens. Taken together, the moral virtues also enable us to fulfill ourselves as human beings. They enable us to attain happiness, by which Aristotle meant self-fulfillment through an active life in accordance with our reason (rather than a life of pleasure or mere contentment).

MacIntyre: Virtues and Practices

Alasdair MacIntyre is a contemporary ethicist who has stimulated a renewed interest in virtue ethics and applied it to thinking about professional ethics.[13] He begins with the idea of *social practices*—cooperative activities aimed toward achieving public goods that could not otherwise be achieved, at least not to the same degree. These *internal goods* are inherent in the practices in that they define what the practices are all about. Thus, the primary internal good of teaching is learning; the internal good of medicine is the promotion of health in accordance with respect for patients' autonomy; and the internal good of engineering is the creation of useful and safe technological products while respecting the autonomy of clients and the public, especially in matters of risk taking. Internal goods differ from *external goods,* like money and prestige, which can be attained through many different kinds of activities and do not define any specific practice.

The virtues especially important to engineering are defined by reference to its internal good. The most basic and comprehensive professional virtue is *professional responsibility,* that is, being morally responsible as a professional. Professional responsibility is an umbrella virtue that encompasses a wide variety of more specific virtues that might be grouped into four categories.

First, *self-direction virtues* are those that are fundamental in exercising moral responsibility.[14] Some of them center on understanding and cognition (as grounded in moral concern): for example, self-understanding and good moral judgment (what Aristotle called practical wisdom). Other self-direction virtues center on commitment and on putting understanding into action: for example, courage, self-discipline, perseverance, fidelity to commitments, self-respect, and integrity. Truthfulness falls under *honesty,* which falls into both groups of self-direction virtues, for it implies truthfulness in cognitive matters and trustworthiness in commitments.

More fully, honesty is a complex virtue. *Honesty in acts* implies respect for the property of others, which includes not stealing, not padding expense sheets, not engaging in bribes and kickbacks. *Honesty in speech* means being candid in revealing all pertinent information, and implies not deceiving, that is, not

---

[13] Alasdair MacIntyre, *After Virtue,* 2nd ed. (Notre Dame, IN: University of Notre Dame Press, 1984). Also see John Kultgen, *Ethics and Professionalism* (Philadelphia: University of Pennsylvania Press, 1988); Albert Flores, ed., *Professional Ideals* (Belmont, CA: Wadsworth, 1988); Michael D. Bayles, *Professional Ethics,* 2nd ed. (Belmont, CA: Wadsworth, 1989).

[14] John Kekes, *The Examined Life* (Lewisburg, PA: Bucknell University Press, 1988).

intentionally misleading others, whether by pretending, manipulating someone's attention, lying (intentionally stating a falsehood in order to mislead), or withholding information that someone has a right to know. *Honesty in beliefs* (intellectual honesty) means forming one's beliefs without self-deception or other forms of evading unpleasant truths and accentuating evidence favorable to one's self-esteem and biases. *Discretion* is sensitivity to the legitimate areas of privacy of the employer or client, especially with regard to confidential information, acknowledging that candid revelation of information is limited by privacy and other moral values.

Second, *public-spirited virtues* are those focused on the good of clients and the wider public affected by one's work. Perhaps most important is justice. The minimum standard is nonmaleficence, that is, not directly and intentionally harming others. Engineering codes of professional conduct also call for beneficence, that is, preventing or removing harm to others by promoting the public safety and welfare. Also important is a sense of community, manifested in faith and hope in the prospects for meaningful life within professional and public communities. Generosity, which means going beyond the minimum requirements in helping others, is shown by engineers who voluntarily give their time, talent, and money to their professional societies and local communities.

Third, *team-work virtues* are those that are especially important in enabling professionals to work successfully with other people. They include collegiality, cooperativeness, loyalty, and respect for legitimate authority. Also important are leadership qualities that play key roles within authority-structured corporations, such as the responsible exercise of authority and the ability to motivate others to meet valuable goals.

Fourth, *proficiency virtues* are the virtues of mastery of one's craft, in particular, mastery of the technical skills that characterize good engineering practice. Following Aristotle, some may prefer to view these virtues as "intellectual virtues" rather than distinctly moral ones. Either way, they are enormously important. The most general proficiency virtue is competence: being well prepared for the jobs one undertakes. Also important is diligence: alertness to dangers and careful attention to detail in performing tasks, for example, by avoiding the deficiency of laziness and the excess of the workaholic. Creativity is especially desirable within a rapidly changing technological society, as is the virtue of self-renewal through continuing education.

Having extended MacIntyre's virtue ethics to engineering, we would suggest that virtue ethics does not actually compete with an ethics of rules, contrary to some advocates of virtue ethics. Obviously, in order for virtues and ideals to be useful in providing guidance, they must make contact with right action. Indeed, can we understand what honesty is in engineering without knowing it implies the truthfulness rule? In general, the basic virtue of professional responsibility implies an array of specific principles such as those articulated in sound codes of ethics. For that reason, we think of virtue ethics as developed by and as complementing theories about right action, rather than competing with theories about right action.

Pragmatism

Pragmatism, which is an approach to ethics developed by the American philosophers William James (1842–1910) and John Dewey (1859–1952), is receiving renewed attention. In some respects, pragmatism is an "anti-theory"—a rejection of the search for general ethical theories. But in other respects, it is a theory about morality that emphasizes the limitations of abstract rules. Either way, it is unfair to confuse it with an appeal to crass expediency, although pragmatists certainly stress the importance of getting (good) results.

Like utilitarians, pragmatists emphasize good consequences, but unlike utilitarians, they embrace a wider range of values than simply maximizing good by impartially considering the

> Moral judgments must remain false and hollow unless they are checked and enlightened by a perpetual reference to the special circumstances that mark the individual lot.
> —George Eliot, *The Mill on the Floss*

interests of everyone affected. Pragmatists emphasize the importance of particular *contexts* in which facts and values must be weighed and balanced, including values such as rights, duties, and virtues. Pragmatists also emphasize *flexibility* in integrating and harmonizing competing values. Rather than applying fixed rules or ideals, moral decision making is essentially a matter of extending moral reasons into new and often uncharted situations.

That is not to say that rules are altogether unimportant. The truthfulness rule and other moral principles can function as important signposts. As Dewey writes:

A moral principle, such as that of chastity, of justice, of the Golden Rule, gives the agent a basis for looking at and examining a particular question that comes up. It holds before him certain possible aspects of the act; it warns him against taking a short or partial view of the act. It economizes his thinking by supplying him with the main heads by reference to which to consider the bearings of his desires and purposes; it guides him in his thinking by suggesting to him the important considerations for which he should be on the lookout.[15]

If rules are to be flexibly applied, so too must rules be revised in light of close attention to cases and contexts. This insistence on attention to cases and context is stressed by Albert Jonsen and Stephen Toulmin's embrace of *casuistry,* which they insist is not sophistry or hair-splitting, but instead attention to paradigm cases of moral and immoral conduct. In place of rules, Jonsen and Toulmin accent the role of analogies, models, extension of past habits, and refined judgment based on experience and social interaction.[16]

Jonsen and Toulmin report that they were led in this direction by their participation in the National Commission for the Protection of Human Subjects of Biomedical and Behavioral Research, from 1975 to 1978. The other members of the Commission were from greatly varied religious, cultural, and political backgrounds, and differed greatly about general moral theory. Nevertheless, they succeeded in reaching substantial agreement on many particular cases, and, emerging from this agreement, recommendations about the moral guidelines for protecting human subjects in medical experiments.

> So long as the debate stayed on the level of particular judgments, the eleven commissioners saw things in much the same way. The moment it soared to the level of "principles," they went their separate ways. Instead of securely established universal principles, in which they had unqualified confidence, giving them intellectual grounding for particular judgments about specific kinds of cases, it was the other way around.[17]

Pragmatism, like act-utilitarianism, carries the danger of paying insufficient attention to moral principles through immersion in specific practical contexts. Nevertheless, in our view, pragmatism is best understood as a methodological emphasis on sensitivity to moral complexity, reasonable compromise, and close attention to the manifold dimensions of cases in their complete context. At least in its most plausible versions, it is not so much an alternative to the search for valid moral principles as it is an

---

[15] John Dewey, *Theory of the Moral Life* (New York: Holt, Rinehart and Winston, 1960), p. 141. First published in 1908. Also see Larry A. Hickman, *John Dewey's Pragmatic Technology* (Bloomington: Indiana University Press, 1990); and William James, "The Moral Philosopher and the Moral Life," in *Essays on Fair and Morals,* ed. Ralph Barton Perry (New York: Meridian Books, 1962), pp. 184–215.

[16] Albert R. Jonsen and Stephen Toulmin, *The Use and Abuse of Casuistry: A History of Moral Reasoning* (Berkeley: University of California Press, 1988).

[17] Ibid., p. 18.

insistence on greater flexibility and tolerance in how principles are applied. As such, one could develop utilitarianism, rights ethics, duty ethics, and virtue ethics in a pragmatic spirit by emphasizing the need to flexibly integrate goods, rights, and virtues within specific situations.

Discussion Topics

1. Kermit Vandivier had worked at B. F. Goodrich for five years, first in instrumentation and later as a data analyst and technical writer. In 1968, he was assigned to write a report on the performance of the Goodrich wheels and brakes commissioned by the Air Force for its new A7-D light attack aircraft. According to his account, he became aware of the design's limitations and of serious irregularities in the qualification tests.[18] The brake failed to meet Air Force specifications. Upon pointing out these problems, however, he was given a direct order to stop complaining and write a report that would show the brake qualified. He was led to believe that several layers of management were behind this demand and would accept whatever distortions might be needed because their engineering judgment assured them the brake was acceptable.

    Vandivier then drafted a 200-page report with dozens of falsifications and misrepresentations. Yet, he refused to sign it. Later he gave as excuses for his complicity the facts that he was 42 years old with a wife and six children. He had recently bought a home and felt financially unable to change jobs. He felt certain that he would have been fired if he had refused to participate in writing the report.

    (a) Assuming for the moment that Vandivier's account of the events is accurate, present and defend your view as to whether Vandivier was justified in writing the report or not.

    (b) Is Vandivier responsible for what he did? In answering this question, distinguish between the various senses of "responsible" discussed in the last chapter.

    (c) Which virtues did Vandivier not display, and what might those virtues have required of him in his situation?

    (d) Vandivier's account of the events has been challenged. After consulting the record of congressional hearings about this case, John Fielder concluded that Vandivier's "claims that the brake was improperly tested and the report falsified are well-supported and convincing, but he overstates the

---

[18] K. Vandivier, "Engineers, Ethics and Economics," in *Conference on Engineering Ethics* (New York: American Society of Civil Engineers, 1975), pp. 20–24.

magnitude of the brake's defects and, consequently, [exaggerates] the danger to the [test] pilot."[19] Comment on the difficulties in achieving clarity when you read only one side of a story and the reputation of the other side may be harmed. Comment also whether the kind of report falsification should be glossed over because the consequences were not serious.

2. Do engineers who work for tobacco companies (for example, in designing cigarette-making machinery) betray their moral integrity, or can they provide an adequate moral accounting for

DOONESBURY © 1989 G. B. Trudeau. Reprinted with permission of UNIVERSAL PRESS SYNDICATE. All rights reserved.

---

[19] John Fielder, "Give Goodrich a Break," *Business and Professional Ethics Journal* 7 (1988), pp. 3–25; Air Force A7-D Brake Problem, Hearing before the Subcommittee on Economy in Government of the Joint Economic Committee, Congress of the United States, Ninety-First Congress, First Session, August 13, 1969. LC card 72-606996.

their work in terms of utilitarianism, rights ethics, duty ethics, Rawls's principles, pragmatism, or some other theory? In your answer, take account of the following arguments (and others of which you may be aware).[20] Tobacco use provides jobs in farming, manufacturing, and retail. Tobacco companies pay taxes, help the balance of payments through export sales, and support charitable causes at home. Tobacco use is responsible for more deaths in the U.S. than alcohol use, drug use, car accidents, homicide, suicide, and AIDS combined, and most new cigarette smokers in the U.S. are teenagers (under 18). There is disagreement of how addictive cigarettes are, but today there are some means of combating smoking habits.

3. Maya d'Angelou has said that courage is the most important virtue (see box on page 52), while Aristotle ranks practical wisdom the highest. Discuss possible reasons for this difference.

4. The following widely discussed case study was written by a leading British philosopher, Bernard Williams. While the case is about a chemist, the issues it raises are equally relevant to engineering. What should George do in order best to preserve his integrity in the situation described below? Would utilitarians and pragmatists offer different answers to whether taking the job betrays integrity? In your answer, discuss whether in taking the job, George would be compromising in either of the two senses of "compromise": (i) undermine integrity by violating one's fundamental moral principles; (ii) settle moral dilemmas and differences by mutual concessions or to reconcile conflicts through adjustments in attitude and conduct.[21]

George, who has just taken his Ph.D. in chemistry, finds it extremely difficult to get a job. He is not very robust in health, which cuts down the number of jobs he might be able to do satisfactorily. His wife has to go out to work to keep [i.e., to support] them, which itself causes a great deal of strain, since they have small children and there are severe problems about looking after them. The results of all this, especially on the children, are damaging. An older chemist, who knows about this situation, says that he can get George a decently paid job in a certain laboratory, which pursues research into chemical and biological warfare. George says that he cannot accept this, since he is opposed to chemical and biological warfare. The older man replies that he is not too keen on it himself, come to that, but after all George's refusal is not going to make the job or the laboratory go away;

[20] Roger Rosenblatt, "How Do Tobacco Executives Live with Themselves?" *New York Times Magazine,* March 20, 1994, pp. 34–41, 55.

[21] Martin Benjamin, *Splitting the Difference: Compromise and Integrity in Ethics and Politics* (Lawrence, KS: University of Kansas Press, 1990).

60-69
#4

what is more, he happens to know that if George refuses the job, it will certainly go to a contemporary of George's who is not inhibited by any such scruples and is likely if appointed to push along the research with greater zeal than George would. Indeed, it is not merely concern for George and his family, but (to speak frankly and in confidence) some alarm about this other man's excess of zeal, which has led the older man to offer to use his influence to get George the job. . . . George's wife, to whom he is deeply attached, has views (the details of which need not concern us) from which it follows that at least there is nothing particularly wrong with research into CBW.[22]

4. A discussion of virtues invites a discussion of vices, or what some would call "sins." Examine the familiar *Seven Deadly Sins* in Roman Catholic doctrine (pride/hubris, avarice, lust, anger, gluttony, envy, sloth) and the *Seven Social Sins* of which Mohandas Gandhi spoke (politics without principle, wealth without work, commerce without morality, pleasure without conscience, education without character, science without humanity, worship without personal sacrifice).

(a) List those that may be encountered in engineering practice, but not necessarily only by engineers.

(b) Discuss the effect, if any, on society and the environment of those "sins" you have listed.

**Customs,
Religion, Self-
Interest, and
Professional
Motives**

We conclude this chapter by examining attempts to reduce moral reasoning to matters of custom, religion, or self-interest. *Ethical relativism* is the view that morality is reducible to custom. *Divine command ethics* says morality is whatever God commands. *Ethical egoism* is the view that morality consists of promoting one's own long-term self-interest. As we proceed, we will identify the genuine moral significance of customs, religion, and self-interest, while suggesting why moral reasoning is more complex than simply an appeal to these things. These topics, especially ethical egoism, will lead us to consider the motivation of engineers and other professionals.

Customs and Ethical Relativism

Ethical relativism is the view that values are relative to and reducible to conventions, customs, or laws. What is right is simply what the customs say is right, and because the customs say it is right. Ethical relativism rules out the possibility of critiquing social customs, laws, and even professional codes of

---

[22] Bernard Williams, *Utilitarianism: For and Against* (New York: Cambridge University Press, 1973), pp. 97–98.

ethics, from a wider moral framework. In particular, if the accepted custom of a group is to lie and deceive whenever it benefits one personally, then the truthfulness rule could not be invoked to argue for a morally better custom that would increase trust and cooperation.

Ethical relativism may at first glance seem to encourage the virtue of tolerance of differences among different groups. Yet, tolerance has limits. Ethical relativism would have us tolerate bribes, cruelty, and intolerance whenever they are customs of a group. Genocide was the customary practice of alarmingly many groups in the twentieth century, but respecting such customs is not a sign of tolerance. Indeed, it implicates us in intolerance.

> [The] relativists' argument is often used by repressive governments to deflect international criticism of their abuse of their citizens. At the very least, anthropologists need to condemn such misuse of cultural relativism. . . .
> —Carolyn Fluehr-Lobban (anthropologist),
> "Cultural Relativism and Universal Rights"

Ethical relativism might seem attractive because it is easily confused with other views that are plausible.[23] One obviously true view is *descriptive relativism,* which is simply the statement that beliefs and attitudes about values differ from culture to culture. Descriptive relativism is obviously true, but it does nothing to establish ethical relativism. For example, the fact that slavery and apartheid were once accepted by some societies does not mean they were justified practices.

Ethical relativism was attractive to early cultural anthropologists, because they tended to overemphasize the extent of moral differences between cultures. Preoccupied with exotic practices such as head-hunting, human sacrifice, and cannibalism, they moved too quickly from "Moral views differ greatly" (descriptive relativism) to "Morality is simply what a culture says it is" (ethical relativism). More recent anthropologists, rethinking their relativism in light of the Holocaust, have drawn attention to underlying similarities between cultural perspectives. They have noted that virtually all cultures show some commitment to promoting social cooperation and to protecting their members against

---

[23] Michael Krausz, ed., *Relativism: Interpretation and Confrontation* (Notre Dame: University of Notre Dame Press, 1989); Michael Stocker, *Plural and Conflicting Values* (Oxford: Clarendon Press, 1990).

needless death and suffering. And beneath moral differences often lie differences in circumstances and in beliefs about facts, rather than differences in moral attitude.

Another obviously true view is what might be called *ethical relationalism,* although it is simply the insistence on paying attention to context: Moral judgments should be made in relation to factors that can vary from case to case. Those factors certainly include customs, conventions, and codes, in addition to an incredible variety of additional facts about the world. Judgments about other cultures have to be based on understanding of and sensitivity to special cultural circumstances. That is because those circumstances are objectively relevant to morality, not because whatever a culture adopts as its laws or customs is automatically justified.

Yet another view easily confused with relativism is *ethical pluralism*: the view that there may be alternative moral perspectives that are reasonable, but no one of which must be accepted completely by all rational and morally concerned persons. It allows that customs can have great moral significance in deciding how we should act. This view is controversial, but we will not challenge that view here; indeed, we think it is true. Moral values are many, varied, and flexible. They permit considerable variation in how different individuals and groups interpret and apply them. Reasonable persons can have reasonable disagreements on moral issues—including issues in engineering ethics.

Religion and Divine Command Ethics

There are important connections between morality and religion in the lives of many people. First, they are related historically. Our moral outlooks today have been shaped in a number of ways by the promulgation of central moral values within major world religions.

Second, for many people, there are important psychological connections between their moral and religious beliefs. For many individuals, religious views support moral responsibility by providing additional motivation for being moral (beyond moral reasons themselves). We are not referring primarily to self-interested motives such as the fear of damnation or other types of punishment, although those motives do play a role for many people. Religious faith and hope imply trust: trust that we can receive insight into what should govern right action and that we can be sustained in that action. Hence, it brings an added inspiration to be moral, even though many people are moral without having religious beliefs.

Third, religions sometimes set a higher moral standard than is conventional. In doing so, many religions emphasize particular ideals of character. For example, the ethics of Christianity centers on the virtues of hope, faith, and especially love; Judaism emphasizes the virtue of *tzedakah* (righteousness); Buddhism emphasizes compassion; Islam emphasizes *ihsan* (translated as either piety or the pursuit of excellence); and Navajo ethics centers on *hozho* (translated variously as harmony, peace of mind, beauty, health, or well-being). To be sure, sometimes sects employ moral standards below what most of us view as acceptable, for instance, by not recognizing the equal rights of women or by treating children in ways that health professionals see as harmful.[24]

What Kant calls the unfathomable metaphysical desire of mankind to find answers to the riddles of the cosmos and the purpose of our life on earth leads some religious groups to explore ways to keep our natural environment sustainable in the face of rising expectations and growing populations. Environmental ethics is discussed in Chapter 6.

Divine command ethics is a doctrine that has occupied philosophers since ancient times. It claims that to say an act is right simply means it is commanded by God, and to say it is wrong means it is forbidden by God. Accordingly, if there were no God to issue commands, then there would be no morality. In this view, morality is reducible to religion, and to disagree morally is essentially to disagree about what God commands. In particular, what truthfulness requires, and even whether it is desirable, would depend entirely on what God commands.

One difficulty raised by this view, of course, is how to know precisely what God's commands are. Another difficulty is knowing whether God exists. (In fact, there are religions not founded on belief in God, such as Confucianism and some versions of Buddhism.) The main difficulty with divine command ethics, however, arises within a framework of belief in a morally good deity. This difficulty pertains to a question asked long ago by Socrates.[25]

Socrates asked, in effect: Why does God make certain commands and not others? Are the commands made on the basis of whim? Surely not, for God is supposed to be morally good and hence would neither approve of nor command such acts as wanton killing, rape, and torture. Divine command ethics has things

---

[24] Margaret P. Battin, *Ethics in the Sanctuary: Examining the Practices of Organized Religion* (New Haven: Yale University Press, 1990); James Davison Hunter, *Culture Wars: The Struggle to Define America* (New York: Basic Books, 1991).

[25] Plato, *Euthyphro,* trans. Lane Cooper, in *The Collected Dialogues of Plato,* ed. Edith Hamilton and Huntington Cairns (Princeton, NJ: Princeton University Press, 1971), pp. 169–85.

backwards. Instead of divine commands creating moral reasons, a morally good deity commands on the basis of moral reasons which kinds of actions are wrong and which are right. For that reason, even most theologians (like St. Thomas Aquinas) have rejected divine command ethics.

Self-Interest and Ethical Egoism

The view that attempts to reduce moral values to self-interest is called *ethical egoism*: *ethical* because it is a theory about morality and *egoism* because it says that the sole duty of each individual is to maximize his or her own good (well-being, happiness). According to its proponents, such as Thomas Hobbes (1588–1679) and the novelist Ayn Rand (1905–1982), moral values are reduced to concern for oneself (prudence), although always a "rational" concern requiring consideration of one's long-term interests.[26] On this view, whether one should be truthful in specific situations or in general turns on whether being so promotes one's well-being.

> Self-love and benevolence, virtue and interest, are not to be opposed, but only to be distinguished from each other . . .
> Every thing is what it is, and not another thing.
> —Joseph Butler, *Fifteen Sermons upon Human Nature*

Defenders of ethical egoism draw a distinction between narrower and wider forms of self-interest. They agree that to be selfishly preoccupied with one's own private good to the point of indifference and disregard of others will generally cut one off from rewarding friendships and love. This leads to the "paradox of happiness": To seek happiness by blinding oneself to other people's happiness leads to one's own unhappiness. Personal well-being generally requires taking some wider interest in others, which in turn implies being truthful with them in order to maintain ties of trust, although the ethical egoist insists that the only reason for showing an interest in others is for the sake of oneself.

Yet, contrary to ethical egoism, friendship and love seem to require—by definition—caring and valuing other people at least in part for their sakes, and not solely for what we can gain from them. So, too, in our view, does professionalism imply direct con-

---

[26] Ayn Rand, *The Virtue of Selfishness* (New York: New American Library, 1964).

cern for moral values. To be sure, usually the requirements of morality and prudence point in the same direction. Thus, the prudent employee and the morally conscientious engineer for the most part look alike in their conduct. It is doubtful, however, that private pursuit of self-interest always works out to everyone's advantage. Morality requires a willingness on the part of both individuals and corporations to place some restraints on self-interest in light of the moral value of other individuals (as all the major ethical theories stress).

These remarks do not constitute a refutation of ethical egoism, and it turns out that ethical egoism cannot be refuted by appealing to moral reasons. That is because ethical egoism amounts to a skeptical rejection of moral reasoning as ordinarily understood. It essentially claims that what are ordinarily viewed as moral reasons (for example, respecting other people's rights or caring about their well-being for their sake) should be disregarded except where they happen to coincide with looking out for one's own neck. According to ethical egoism, "Number 1" is all that counts. Such a view denies the validity of moral reasons.

To be sure, any adequate ethical theory will appreciate the importance of self-interest. Just as other people have moral worth, so do I, and surely each of us has a special responsibility for pursuing our self-fulfillment—for example, through finding meaningful work. Thus, utilitarians emphasize that one's own good is as valuable as that of anyone else. Duty ethicists emphasize the importance of duties to oneself, especially the duties to be prudent and to maintain self-respect, and these duties imply due regard for one's overall well-being. Rights ethicists begin with the idea of rights to pursue one's legitimate interests. And virtue ethics is greatly concerned with the role of the virtues in furthering self-fulfillment.

Psychological Egoism

Ethical egoism may seem attractive if one believes that the people are psychologically incapable of caring about anyone but themselves. This view is called *psychological egoism*: All people are always and only motivated by what they believe is good for themselves in some respect. Note that this is a claim about the facts, about what actually motivates persons, rather than (like ethical egoism) a moral theory about what should motivate them. It is a claim about what *is* true concerning human nature (as a matter of descriptive or factual inquiry), rather than a theory about what *ought* to be. Nevertheless, if psychological egoism were true, it would provide support for ethical egoism. If the only thing that we can care about is ourselves, then it would seem that we should at least adopt an enlightened and long-term view of our self-interest.

But is self-seeking the only human motive? Are we universally, "always and only," concerned only about ourselves? No doubt, self-seeking is a very strong motive, indeed usually the strongest one. That view is called *predominant egoism*: The strongest desire for most people most of the time is self-seeking.[27] Such a view is plausible and open to scientific confirmation. It is also plausible to believe that most acts of helping and service to others involve *mixed motives,* that is, a combination of self-concern and concern for others. But the psychological egoist's reduction of all human motives to self-seeking seems highly implausible. It seems to fly in the face of facts about human capacities for love, friendship, community involvement—and professional commitments to the public good. In general, it denies that we can care about moral values—rights, duties, goods, virtues—in their own right, rather than solely for how they serve our own purposes.

Why, then, have some people, including many distinguished psychologists, believed psychological egoism? Perhaps because of a group of seductive confusions. Here are three such confusions and arguments built on them.

First, it is contended, people always act on their own desires; therefore, they always and only seek something for themselves, namely the satisfaction of their desires. In reply, surely there are many different kinds of desires, depending on their objects—what the desire is for. When we desire goods for ourselves, we are self-seeking; but when we desire goods for other people (for their sake), we are not. The mere fact that in both instances we act on our own desires does nothing to support psychological egoism.

Second, people always seek pleasures; therefore, they always and only seek something for themselves, namely their pleasures. In reply, there are different sources of pleasures. Taking pleasure in seeking and getting a good solely for oneself is different from taking pleasure in helping others.

Third, we can always imagine there is an ulterior, self-seeking motive present whenever a person helps someone else; therefore, people always and only seek goods for themselves. In reply, there is a difference between imagination and reality! Also, the mere presence of an element of self-seeking does not establish that self-seeking is the sole motive.

If we avoid these confusions that make psychological egoism seem plausible, we open the door to seeing professionals as capable of genuinely caring about the good of clients, colleagues, and the community. Of course, professionals, as indeed all persons,

---

[27] Gregory S. Kavka, *Hobbesian Moral and Political Theory* (Princeton, NJ: Princeton University Press, 1986).

are strongly motivated by self-interest, but they are also capable of responding to moral reasons in their own right, as well as additional motives concerned with the particular nature of their work. Let us develop this point by considering the range of motives that typically sustain professional commitments.

Meaningful Work and Professionals' Commitments

Most persons are drawn to professions because they constitute especially meaningful forms of work. Meaningful work has value and strongly motivates participants by providing a sense of accomplishment and self-worth. We can distinguish, however, a variety of values, commitments, and sources of motivation that provide meaning. Many of them cluster into three categories: craft, moral caring, and compensation.

Craft values and motives center on excellence in meeting the technical standards of a profession, together with related aesthetic values of beauty. The values of *moral caring* and its motives center on providing valuable services to the community, as well as caring relationships among professionals, other involved workers, and clients. *Compensation* values and motives are for social rewards such as money, power, recognition, and job or career stability. In practice, these motives are interwoven and usually mutually supportive, but let us separate them out for further comment. In doing so, we will attempt to further clarify what moral values are by comparing and contrasting them with other types of values.

Craft Motives

The undergraduate curriculum for engineering is generally acknowledged to be more rigorous and difficult than the majority of academic disciplines. We might guess that students are attracted to engineering at least in part because of the challenge it offers to intelligent people. Do empirical studies back up this somewhat flattering portrayal? To a significant extent, yes. Typically, students are motivated to enter engineering primarily by a desire for interesting and challenging work. They have an "activist orientation" in the sense of wanting to create concrete objects and systems—to build them and to make them work. They are more skilled in math than average college students, although they tend to have a low tolerance for ambiguities and uncertainties that cannot be measured and translated into figures.[28]

---

[28] Robert Perrucci and Joel E. Gerstl, *Profession without Community: Engineers in American Society* (New York: Random House, 1969), pp. 27–52.

What is it that is so appealing and challenging in making technological products? Perhaps no one has so elegantly conveyed the excitement of engineering as Samuel Florman in his book *The Existential Pleasures of Engineering*.[29] By "existential pleasures," Florman means deep-rooted and elemental satisfactions, such as the following five sources of enjoyment.

First, the desire to improve the world engages individuals' sense of personal involvement and power. Second, the joy of practical and creative effort includes planning, designing, testing, producing, selling, constructing, and maintaining. There is an accompanying pride in achieving excellence in the technical aspects of one's work. Third, engineers share with scientists moments of peace and wonder in understanding the world. Fourth, the sheer magnitude of natural phenomena—oceans, rivers, mountains, and prairies—both inspires and challenges undertakings of enormous scale in conceiving immense ships, bridges, tunnels, communication links, and other vast undertakings. Equally challenging now is the world of microsystems. Fifth, engineers live in the presence of machines that can generate a comforting and absorbing sense of a manageable, controlled, and ordered world.

Moral Motives

As Florman also points out, most engineers also live with a sense of helping, of contributing to the well-being of other human beings. That is, they have a sense of caring and respect for persons. This needs to be borne in mind when interpreting empirical studies of students' main motives for entering engineering. The technical challenge it offers is always seen by them against the background of participating in a socially useful and important enterprise. Engineering would not attract students if it were viewed by them as generally directed toward immoral ends.

Emphasizing the idea of morality as helping, as concern for the good of others as well as oneself, serves to counterbalance the frequently negative view of morality as a set of onerous constraints and unpleasant disagreements. Too often, the mention of ethics evokes groans, rather than engagement, because it brings to mind occasions in which individuals find themselves in disputes and even animosity. Indeed, as we noted earlier, ethics can involve several different kinds of conflicts and disputes, and it sometimes does place firm obligations on us. Nevertheless, it is important to draw attention to the positive moral ideals that contribute to professional life.

---

[29] Samuel C. Florman, *The Existential Pleasures of Engineering* (New York: St. Martin's, 1976), pp. 120–21.

Compensation and Self-Interest

Compensation values—such as money, power, and recognition—can be pursued for the good of others, but even in promoting one's long-term well-being, they are genuine goods and play a major role in motivating and guiding human conduct. Indeed, reasonable regard for one's self-interest is a moral virtue—the virtue of *prudence*—as long as it does not crowd out other virtues.

To conclude, professionals are motivated by many different reasons falling into the categories of compensation, craft, religion, and moral caring. All these motives are sources of satisfaction and thus contribute to self-interest, but the "self" whose interest is at issue is far more complex than psychological egoism suggests.

Discussion Topics

1. With regard to each of the following cases, first discuss what morality requires and then what self-interest requires. Is the answer the same or different?

   (a) Bill, a process engineer, learns from a former classmate, who is now an Occupational Safety and Health Administration (OSHA) regional compliance officer, that there will be an unannounced inspection of Bill's plant. Bill believes that unsafe practices are often tolerated in the plant, especially in the handling of toxic chemicals. Although there have been small spills, no serious accidents have occurred in the plant during the past few years. What should Bill do?[30]

   (b) On a midnight shift, a botched solution of sodium cyanide, a reactant in an organic synthesis, is temporarily stored in drums for reprocessing. Two weeks later, the day shift foreperson cannot find the drums. Roy, the plant manager, finds out that the batch has been illegally dumped into the sanitary sewer. He severely disciplines the night shift foreperson. Upon making discreet inquiries, he finds out that no apparent harm has resulted from the dumping.[31] Should Roy inform government authorities, as is required by law in this kind of situation?

2. Long before H. G. Wells wrote *The Invisible Man,* Plato (428–348 B.C.) in *The Republic* described a shepherd named Gyges who, according to a Greek legend, discovers a ring that enables him to become invisible when he turns its bezel. Gyges uses his magical powers to seduce the queen, kill the king, and

---

[30] Jay Matley, Richard Greene, and Celeste McCauley, "Health, Safety and Environment," *Chemical Engineering* 28 (September 1987), p. 115.
[31] Ibid., p. 117.

take over an empire. If we have similar powers, why should we feel bound by moral constraints? In particular, if professionals are sufficiently powerful to pursue their desires without being caught for malfeasance, why should they care about the good of the wider public?

In your answer, reflect on the question "Why be moral?" Is the question asking for self-interested reasons for being moral, and, if so, does it already presuppose that only self-interest, not morality, provides valid reasons for conduct?

3. What problems, if any, do you see in invoking religious views, especially those based in particular religious literature, in discussing ethical issues in engineering—both in the classroom and inside the corporation? Even if individuals ground their moral views in their religious convictions, does professionalism imply being able to use distinctively moral (rather than religious) concepts in discussing issues in professional ethics?

4. A *work ethic* is a set of attitudes, which implies a motivational orientation, concerning the value of work.[32] Which, if any, of the following work ethics do you find attractive, and why? Which of them, as applied to engineering, are compatible or incompatible with the kinds of commitments desirable for professionals?

   (a) The Protestant work ethic, as named and analyzed by sociologist Max Weber in *The Protestant Ethic and the Spirit of Capitalism,* was the idea that financial success is a sign that predestination has ordained one as favored by God. This was thought to imply that making big profits is a duty mandated by God. Profit becomes an end in itself rather than a means to other ends. It is to be sought rationally, diligently, and perhaps without compromise with other values such as spending time with one's family.

   (b) Work is a necessary evil. It is the sort of thing one must do in order to avoid worse evils, such as dependency and poverty. But it is mind-numbing, degrading, and a major source of anxiety and unhappiness.

   (c) Work is the major instrumental good in life. It is the central means for providing the income needed to avoid economic dependence on others, for obtaining desired goods and services, and for achieving status and recognition from others.

   (d) Work is intrinsically valuable to the extent that it is enjoyable or meaningful in allowing personal expression and self-fulfillment. Meaningful work is worth doing for its own sake and for the sense of personal identity and self-esteem it brings.

---

[32] David J. Cherrington, *The Work Ethic* (New York: AMACOM, 1980), pp. 19–30, 253–74.

[Handwritten margin notes:]

To Dr. M. Flynn S M.D
From Justin Allen
Date: Feb. 13, 2004
Subject: Personal Work ethics

- enjoy working
- feel satisfied
- job well done
- I like to accomplish things
- do the best I can
- Make others feel good
- when others are satisfied I am too
- always to be the best no matter what.
- like to see results once I get going

3

Engineering as Social Experimentation

As it departed on its maiden voyage in April 1912, the *Titanic* was proclaimed the greatest engineering achievement ever. Not merely was it the largest ship the world had seen, having a length of two and a half football fields, it was also the most glamorous of ocean liners, complete with a tropical vinegarden restaurant and the first seagoing masseuse. It was supposed to be the first totally safe ship. Since the worst collision envisaged was at the juncture of 2 of its 16 watertight compartments, and since it could float with any four compartments flooded, the *Titanic* was confidently believed to be virtually unsinkable.

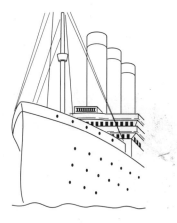

Buoyed by such confidence, the captain allowed the ship to sail full speed at night in an area frequented by icebergs, causing a collision which tore a large gap in the ship's side, directly or indirectly flooding five compartments. Time remained to evacuate the ship, but there were not enough lifeboats to accommodate all the passengers and crew. British regulations then in effect did not foresee vessels of this size. Accordingly, only 825 places were required in lifeboats, sufficient for a mere one-quarter of the *Titanic*'s capacity of 3547 passengers and crew. No extra precautions had seemed necessary for a ship believed to be practically unsinkable. The result: 1522 dead (drowned or frozen) out of the 2227 on board for the Titanic's first trip.[1]

The *Titanic* remains a haunting image of technological complacency. Most products of technology present some potential dangers, and thus engineering is an inherently risky activity. In order to underscore this fact and help in exploring its ethical

---

[1] Walter Lord, *A Night to Remember* (New York: Holt, 1976); Wynn C. Wade, *The Titanic: End of a Dream* (New York: Penguin, 1980); Michael Davie, *The Titanic* (London: The Bodley Head, 1986).

implications, we suggest that engineering should be viewed as an experimental process. It is not, of course, an experiment conducted solely in a laboratory under controlled conditions. Rather, it is an experiment on a social scale involving human subjects.

Wherever great risk to human life is involved, a ready means of escape (a "safe exit") should be provided. With this in mind, it should not matter why the *Titanic* sank. There are conjectures that the *Titanic* left England with a coal fire on board, that this made the captain rush the ship to New York, and that water entering the coal bunkers through the gash caused an explosion and greater damage to the compartments. Others maintain that embrittlement of the ship's steel hull in the icy waters caused a much larger crack than a collision would otherwise have produced. Shipbuilders have argued that having the watertight bulkheads reach higher on such a big ship would have kept the ship afloat, but this would have restricted space on the passenger decks for cabins and paying passengers. However, what matters most is that the lack of lifeboats and the difficulty of launching those available from the listing ship prevented safe exit for most persons aboard.

Engineering as Experimentation

Experimentation is commonly recognized to play an essential role in the design process. Preliminary tests or simulations are conducted from the time a new design concept is given its first rough design. Materials and processes are tried out, usually employing formal experimental techniques. Such tests serve as the basis for more detailed designs, which in turn are tested. At the production stage, further tests are run, until a finished product evolves. The normal design process is thus iterative, carried out on trial designs with modifications being made on the basis of feedback information acquired from tests. Beyond those specific tests and experiments, however, each engineering project taken as a totality may itself be viewed as an experiment.

Similarities to Standard Experiments

Several features of virtually every kind of engineering practice combine to make it appropriate to view engineering projects as experiments. First, any project is carried out in partial ignorance. There are uncertainties in the abstract model used for the design calculations; there are uncertainties in the precise characteristics of the materials purchased; there are uncertainties in the precision of materials processing and fabrication; there are uncertainties about the nature of the stresses the finished product will encounter. Engineers do not have the luxury of waiting until all the relevant facts are in before commencing work. At some point, theoretical exploration and laboratory testing must be bypassed for the sake of moving ahead on a project. Indeed, one talent crucial to an engineer's success lies precisely in the ability to accomplish tasks safely with only a partial knowledge of scientific laws about nature and society.

> To undertake a great work, and especially a work of a novel type, means carrying out an experiment. It means taking up a struggle with the forces of nature without the assurance of emerging as the victor after the first attack.
> —Louis Marie Henri Navier (1785–1836),
> bridge builder, founder of structural analysis

Second, the final outcomes of engineering projects, like those of experiments, are generally uncertain. Often in engineering, it is not even known what the possible outcomes may be, and great risks may attend even seemingly benign projects. A reservoir may do damage to a region's social fabric or to its ecosystem. It may not even serve its intended purpose if the dam leaks or breaks. An aqueduct in a region where it is the only source of water, may bring about a population explosion, creating dependency and vulnerability without adequate safeguards. A jumbo airplane may bankrupt the small airline that bought it as a status symbol. A special-purpose fingerprint reader may find its main application in the identification and surveillance of dissidents by totalitarian regimes. A nuclear reactor, the scaled-up version of a successful smaller model, may exhibit unexpected problems that endanger the surrounding population, leading to its untimely shutdown at great cost to owner and consumers alike. In the past, a hair dryer may have exposed the unwary user to lung damage from the asbestos insulation in its barrel.

Third, effective engineering relies on knowledge gained about products both before and after they leave the factory—knowledge needed for improving current products and creating better ones. That is, ongoing success in engineering depends upon gaining new knowledge, just as does ongoing success in experimentation. Monitoring is thus as essential to engineering as it is to experimentation in general. To monitor is to make periodic observations and tests in order to check for both successful performance and unintended side effects. But since the ultimate test of a product's efficiency, safety, cost-effectiveness, environmental impact, and aesthetic value lies in how well that product functions within society, monitoring cannot be restricted to the in-house development or testing phases of an engineering venture. Monitoring should also extend to the stage of client use, because just as in experimentation, both the intermediate and final results of an engineering project deserve analysis if the correct lessons are to be learned from it.

Learning from the Past

It might be expected that engineers would learn not only from their own earlier design and operating results, but also from those of other engineers. Unfortunately, that is frequently not the case. Lack of established channels of communication, misplaced pride in not asking for information, embarrassment at failure or fear of litigation, and plain neglect often impede the flow of such information and lead to many repetitions of past mistakes. Here are a few examples:

1. The *Titanic* lacked a sufficient number of lifeboats decades after most of the passengers and crew on the steamship *Arctic* had perished because of the same problem.[2]

2. "Complete lack of protection against impact by shipping caused Sweden's worst ever bridge collapse on Friday as a result of which eight people were killed." Thus reported the *New Civil Engineer* on January 24, 1980. On May 15 of the same year, it also reported the following: "Last Friday's disaster at Tampa Bay, Florida, was the largest and most tragic of a growing number of incidents of errant ships colliding with bridges over navigable waterways." While collisions of ships with bridges do occur— other well-known cases being those of the Maracaibo Bridge (Venezuela, 1964) and the Tasman Bridge (Australia, 1975)— Tampa's Sunshine Skyline Bridge was not designed with horizontal impact forces in mind because the code did not require them. Some engineers have proposed the use of floating concrete bumpers that can deflect ships.

3. In June 1966, a section of the Milford Haven bridge in Wales collapsed during construction. In October of the same year, a bridge of similar design was being erected by the same bridgebuilder (Freeman Fox and Partners) in Melbourne, Australia, when it too partially collapsed, killing 33 people and injuring 19. This happened shortly after chief construction engineer Jack Hindshaw (also a casualty) had assured worried workers that the construction site was safe.[3]

4. Valves are notorious for being among the least reliable components of hydraulic systems. It was a pressure relief valve, and lack of definitive information regarding its open or shut state, that contributed to the nuclear reactor accident at Three Mile Island

---

[2] Wade, *The Titanic,* p. 417.
[3] "Yarrow Bridge," editorial, *The Engineer* 210 (October 1970), p. 415.

on March 28, 1979. Similar malfunctions had occurred with identical valves on nuclear reactors at other locations. The required reports had been filed with Babcock and Wilcox, the reactor's manufacturer, but no attention had been given to them.[4]

These examples, and others to be given in later chapters, illustrate why it is not sufficient for engineers to rely on handbooks and computer programs without knowing the limits of the tables and algorithms underlying their favorite tools. They need to visit shop floors and construction sites to learn from workers and foremen how earlier projects have fared during erection or assembly and tests, and how satisfied the customers were. The art of back-of-the-envelope calculations to obtain ball-park values with which to independently check more lengthy and complicated procedures must not be lost. Engineering, just like experimentation, demands practitioners who remain alert and well-informed at every stage of a project's history, and who exchange ideas freely with colleagues in related departments.

Contrasts with Standard Experiments

To be sure, engineering differs in some respects from standard experimentation. Some of those very differences help to highlight the engineer's special responsibilities. And exploring the differences can also aid our thinking about the moral responsibilities of all those engaged in engineering.

Experimental Control

One great difference arises with experimental control. In a standard experiment, this involves the selection, at random, of members for two different groups. The members of one group receive the special, experimental treatment. Members of the other group, called the control group, do not receive that special treatment, although they are subjected to the same environment as the first group in every other respect.

In engineering, this is not the usual practice—unless the project is confined to laboratory experimentation—because the experimental subjects are human beings out of the experimenter's control. Indeed, clients and consumers exercise most of the control because it is they who choose the product or item they wish to use. This makes it impossible to obtain a random selection of participants from various groups. Nor can parallel control groups be established based on random sampling. Thus, it is not possible to study the effects that changes in variables have on

---

[4] Robert Sugarman, "Nuclear Power and the Public Risk," *IEEE Spectrum* 16 (November 1979), p. 72.

two or more comparison groups, and one must simply work with the available historical and retrospective data about various groups that use the product.

This suggests that the view of engineering as social experimentation involves a somewhat extended usage of the concept of experimentation. Nevertheless, "engineering as social experimentation" should not be dismissed as a merely metaphorical notion. There are other fields where it is not uncommon to speak of experiments whose original purpose was not experimental in nature and that involve no control groups.

For example, social scientists monitor and collect data on differences and similarities between existing educational systems that were not initially set up as systematic experiments. In doing so, they regard the current diversity of systems as constituting what has been called a "natural experiment" (as opposed to a deliberately initiated one).[5] Similarly, we think that engineering can be appropriately viewed as just such a "natural experiment" using human subjects, despite the fact that most engineers do not currently consider it in that light.

Informed Consent

Viewing engineering as an experiment on a societal scale places the focus where it should be: on the human beings affected by technology, for the experiment is performed on persons, not on inanimate objects. In this respect, albeit on a much larger scale, engineering closely parallels medical testing of new drugs and techniques on human subjects.

Society has recently come to recognize the primacy of the subject's safety and freedom of choice as to whether to participate in medical experiments. Ever since the revelations of prison and concentration camp horrors in the name of medicine, an increasing number of moral and legal safeguards have arisen to ensure that subjects in experiments participate on the basis of informed consent.

While current medical practice has increasingly tended to accept as fundamental the subject's moral and legal rights to give informed consent before participating in an experiment, contemporary engineering practice is only beginning to recognize those rights. We believe that the problem of informed consent, which is so vital to the concept of a properly conducted experiment involving human subjects, should be the keystone in the interaction between engineers and the public. We are talking about the lay public. When a manufacturer sells a new device to

---

[5] Alice M. Rivlin, *Systematic Thinking for Social Action* (Washington, DC: The Brookings Institution, 1971), p. 70.

a knowledgeable firm that has its own engineering staff, there is usually an agreement regarding the shared risks and benefits of trying out the technological innovation.

Informed consent is understood as including two main elements: knowledge and voluntariness. First, subjects should be given not only the information they request, but all the information needed to make a reasonable decision. Second, subjects must enter into the experiment without being subjected to force, fraud, or deception. Respect for the fundamental rights of dissenting minorities and compensation for harmful effects are taken for granted here.

The mere purchase of a product does not constitute informed consent, any more than does the act of showing up on the occasion of a medical examination. The public and clients must be given information about the practical risks and benefits of the process or product in terms they can understand. Supplying complete information is neither necessary nor in most cases possible. In both medicine and engineering, there may be an enormous gap between the experimenter's and the subject's understanding of the complexities of an experiment. But while this gap most likely cannot be closed, it should be possible to convey all pertinent information needed for making a reasonable decision on whether to participate.

We do not propose a proliferation of lengthy environmental impact reports. We favor the kind of sound advice a responsible physician gives a patient when prescribing a course of drug treatment that has possible side effects. The physician must search beyond the typical sales brochures from drug manufacturers for adequate information; hospital management must allow the physician the freedom to undertake different treatments for different patients, as each case may constitute a different "experiment" involving different circumstances; finally, the patient must be readied to receive the information.

Likewise, an engineer cannot succeed in providing essential information about a project or product unless there is cooperation by management and also receptivity on the part of those who should have the information. Management is often understandably reluctant to provide more information than current laws require, fearing disclosure to potential competitors and exposure to potential lawsuits. Moreover, it is possible that, paralleling the experience in medicine, clients or the public may not be interested in all of the relevant information about an engineering project, at least not until a crisis looms. It is important nevertheless that all avenues for disseminating such information be kept open and ready.

We note that the matter of informed consent is surfacing indirectly in the continuing debate over acceptable forms of energy.

Representatives of the nuclear industry can be heard expressing their impatience with critics who worry about reactor malfunction while engaging in statistically more hazardous activities such as driving automobiles and smoking cigarettes. But what is being overlooked by those industry representatives is the common enough human readiness to accept *voluntarily undertaken risks* (as in daring sports), even while objecting to *involuntary risks* resulting from activities in which the individual is neither a direct participant nor a decision maker. In other words, we all prefer to be the subjects of our own experiments rather than those of somebody else. When it comes to approving a nearby oil-drilling platform or a nuclear plant, affected parties expect their consent to be sought no less than it is when a doctor contemplates surgery.

Prior consultation of the kind suggested can be effective. When Northern States Power Company (Minnesota) was planning a new power plant, it got in touch with local citizens and environmental groups before it committed large sums of money to preliminary design studies. The company was able to present convincing evidence regarding the need for a new plant and then suggested several sites. Citizen groups responded with a site proposal of their own. The latter was found acceptable by the company. Thus, informed consent was sought from and voluntarily given by those the project affected, and the acrimonious and protracted battles so common in other cases where a company has already invested heavily in decisions based on engineering studies alone was avoided.[6] Note that the utility company interacted with groups that could serve as proxy for various segments of the rate-paying public. Obviously it would have been difficult to involve the rate-payers individually.

We endorse a broad notion of informed consent, or what some would call *valid consent,* defined by the following conditions:[7]

1. The consent was given voluntarily.
2. The consent was based on the information that a rational person would want, together with any other information requested, presented in understandable form.
3. The consenter was competent (not too young or mentally ill, for instance) to process the information and make rational decisions.

---

[6] Peter Borrelli, Mahlon Easterling, Burton H. Klein et al., *People, Power and Pollution* (Pasadena, CA: Environmental Quality Lab, California Institute of Technology, 1971), pp. 36–39.

[7] Charles M. Culver and Bernard Gert, "Valid Consent," in *Conceptual and Ethical Problems in Medicine and Psychiatry,* ed. Charles M. Culver and Bernard Gert (New York: Oxford University Press, 1982).

We suggest two requirements for situations in which the subject cannot be readily identified as an individual:

4. Information that a rational person would need, stated in understandable form, has been widely disseminated.

5. The subject's consent was offered in proxy by a group that collectively represents many subjects of like interests, concerns, and exposure to risk.

Knowledge Gained

Scientific experiments are conducted to gain new knowledge, while "engineering projects are experiments that are not necessarily designed to produce very much knowledge," according to a valuable interpretation of our paradigm by Taft Broome.[8] When we carry out an engineering activity as if it were an experiment, we are primarily preparing ourselves for unexpected outcomes. The best outcome in this sense is one that tells us nothing new but merely affirms that we are right about something. Unexpected outcomes send us on a search for new knowledge—possibly involving an experiment of the first (scientific) type. For the purposes of our model, the distinction is not vital because we are concerned about the manner in which the experiment is conducted, such as that valid consent of human subjects is sought, safety measures are taken, and means exist for terminating the experiment at any time with all participants having access to a safe exit.

Discussion Topics

1. On June 5, 1976, Idaho's Teton Dam collapsed, killing 11 people and causing $400 million in damage. The Bureau of Reclamation, which built the ill-fated Teton Dam, allowed it to be filled rapidly, thus failing to provide sufficient time to monitor for the presence of leaks in a project constructed with less-than-ideal soil.[9]

   Drawing upon the concept of engineering as social experimentation, discuss the following facts uncovered by the General Accounting Office and reported in the press.

   (a) Because of the designers' confidence in the basic design of Teton Dam, it was believed that no significant water seepage would occur. Thus, sufficient instrumentation to detect water erosion was not installed.

---

[8] Taft H. Broome Jr., "Engineering Responsibility for Hazardous Technologies," *Journal of Professional Issues in Engineering* 113 (April 1987), pp. 139–49.

[9] Gaylord Shaw, "Bureau of Reclamation Harshly Criticized in New Report on Teton Dam Collapse," *Los Angeles Times,* June 4, 1977, Part I, p. 3; Philip M. Boffey, "Teton Dam Verdict: Foul-up by the Engineers," *Science* 195 (January 1977), pp. 270–72.

(b) Significant information suggesting the possibility of water seepage was acquired at the dam site six weeks before the collapse. It was sent through routine channels from the project supervisors to the designers, and arrived at the designers the day after the collapse.

(c) During the important stage of filling the reservoir, there was no around-the-clock observation of the dam. As a result, the leak was detected only five hours before the collapse. Even then, the main outlet could not be opened to prevent the collapse because a contractor was behind schedule in completing the outlet structure.

(d) Ten years earlier, the Bureau's Fontenelle Dam had experienced massive leaks that caused a partial collapse, an experience on which the Bureau could have drawn.

2. Read about the catastrophic failure of a dam (other than the Teton Dam), then prepare a list of procedural questions that designers, builders, and operators of dams as well as local public safety officers and nearby neighbors of dams should be able to answer. (You may wish to consult *Why Buildings Fail* by M. Levy and M. Salvadori [New York: W. W. Norton & Co., 1987] for cases such as the 1889 Johnstown flood in Pennsylvania and the 1959 collapse of Malpasset Dam in France. Another good source is *Dams and Public Safety* by R. B. Jansen [Denver: USID, Government Printing Office, 1980].)

3. The University of California uses tax dollars to develop farm machinery such as tomato, lettuce, and melon harvesters and fruit tree shakers. Such machinery reduces the need for farm labor and raises farm productivity. It definitely benefits the growers. It is also said to benefit all of society. Farm workers, however, claim that replacing an adequate and willing workforce with machines will generate social costs not offset by higher productivity. Among the costs they cite are the need to retrain farm workers for other jobs and the loss of small farms. Discuss if and how continuing farm mechanization may be viewed as an experiment.

4. Models often influence thinking by effectively organizing and guiding reflection and crystallizing attitudes. Yet they usually have limitations and can themselves be misleading to some degree. Write a short essay in which you critically assess the strengths and weaknesses you see in the social experimentation model.

One possible criticism you might consider is whether the model focuses too much on the creation of new products, whereas a great deal of engineering involves the routine application of results from past work and projects. Another point to consider is how informed consent is to be measured in situations where groups are involved, as in the construction of a garbage incinerator near a community of people having mixed views about the advisability of constructing the incinerator.

5. Debates over responsibility for safety in regard to technological products often turn on who should be considered mainly responsible: the consumer ("buyer beware") or the manufacturer ("seller beware"). How might an emphasis on the idea of informed consent influence thinking about this question?

6. During 1994, the U.S. government released information that in the decades following World War II, some of its agencies conducted tests on the effects of radiation on unsuspecting civilians. Discuss such practices in the light of secrecy imposed by national security considerations and of the right of subjects of experimentation to be informed of the nature of the tests and their possible effects.

Engineers as Responsible Experimenters

What are the responsibilities of engineers to society? Viewing engineering as social experimentation does not by itself answer this question. While engineers are the main technical enablers or facilitators, they are far from being the sole experimenters. Their responsibility is shared with management, the public, and others. Yet their expertise places them in a unique position to monitor projects, to identify risks, and to provide clients and the public with the information needed to make reasonable decisions. We want to know what is involved in displaying the virtue of being a responsible person while acting as an engineer. From the perspective of engineering as social experimentation, what are the general features of morally responsible engineers?

At least four elements are pertinent—a conscientious commitment to live by moral values, a comprehensive perspective, autonomy, and accountability[10]—or, stated in greater detail as applied to engineering projects conceived as social experiments:

1. A primary obligation to protect the safety of human subjects and respect their right of consent.

2. A constant awareness of the experimental nature of any project, imaginative forecasting of its possible side effects, and a reasonable effort to monitor them.

3. Autonomous, personal involvement in all steps of a project.

4. Accepting accountability for the results of a project.

---

[10] Graham Haydon, "On Being Responsible," *The Philosophical Quarterly* 28 (1978), pp. 46–57.

It is implied in the foregoing that engineers should also display technical competence and other attributes of professionalism. Inclusion of these four requirements as part of engineering practice would then earmark a definite "style" of engineering. In elaborating upon this style, we will note some of the contemporary threats to it.

Conscientiousness

People act responsibly to the extent that they conscientiously commit themselves to live according to moral values. But moving beyond this truism leads immediately to controversy over the precise nature of those values. Moral values transcend a consuming preoccupation with a narrowly conceived self-interest. Accordingly, individuals who think solely of their own good to the exclusion of the good of others are not moral agents. By conscientious moral commitment we mean a sensitivity to the full range of moral values and responsibilities relevant to a given situation, and the willingness to develop the skill and expend the effort needed to reach the best balance possible among those considerations. It will be noted that conscientiousness implies consciousness (in the sense of awareness), because intent is not sufficient. Open eyes, open ears, and an open mind are required to recognize a given situation, its implications, and who is involved or affected.

The contemporary working conditions of engineers tend to narrow moral vision solely to the obligations that accompany employee status. As stated earlier, some 90 percent of engineers are salaried employees, most of whom work within large bureaucracies under great pressure to function smoothly within the organization. There are obvious benefits in terms of prudent self-interest and concern for one's family that make it easy to emphasize as primary the obligations to one's employer. Gradually, the minimal negative duties, such as not falsifying data, not violating patent rights, and not breaching confidentiality, may come to be viewed as the full extent of moral aspiration.

Conceiving engineering as social experimentation restores the vision of engineers as guardians of the public interest, whose professional duty it is to guard the welfare and safety of those affected by engineering projects. And this helps to ensure that such safety and welfare will not be disregarded in the quest for new knowledge, the rush for profits, a narrow adherence to rules, or a concern over benefits for the many that ignores harm to the few.

The role of social guardian should not suggest that engineers force, paternalistically, their own views of the social good upon society. For, as with medical experimentation on humans, the social experimentation involved in engineering should be restricted by the participant's consent—voluntary and informed consent.

Relevant Information

Conscientiousness is blind without relevant factual information. Hence, showing moral concern involves a commitment to obtain and properly assess all available information pertinent to meeting one's moral obligations. This means, as a first step, fully grasping the context of one's work, which makes it count as an activity having a moral import.

For example, there is nothing wrong in itself with trying to design a good heat exchanger. But if I ignore the fact that the heat exchanger will be used in the manufacture of a potent, illegal hallucinogen, I am showing a lack of moral concern. It is this requirement that one be aware of the wider implications of one's work that makes participation in, say, a design project for a super-weapon morally problematic—and that makes it sometimes convenient for engineers self-deceivingly to ignore the wider context of their activities, a context that may rest uneasily with an active conscience.

Another way of blurring the context of one's work results from the ever-increasing specialization and division of labor that makes it easy to think of someone else in the organization as responsible for what otherwise might be a bothersome personal problem. For example, a company may produce items with obsolescence built into them, or the items might promote unnecessary energy usage. It is easy to place the burden on the sales department: "Let them inform the customers—if the customers ask." It may be natural to thus rationalize one's neglect of safety or cost considerations, but it shows no moral concern. More convenient is a shifting of the burden to the government and voters: "We will attend to this when the government sets standards so our competitors must follow suit," or "Let the voters decide on the use of super-weapons; we just built them."

These ways of losing perspective on the nature of one's work also hinder one in acquiring a full perspective along a second dimension of factual information: the consequences of what one does. And so, while regarding engineering as social experimentation points out the importance of context, it also urges the engineer to view his or her specialized activities in a project as part of a larger whole having a social impact—an impact that may involve a variety of unintended effects. Accordingly, it emphasizes the need for wide training in disciplines related to engineering and its results, as well as the need for a constant effort to imaginatively foresee dangers.

It might be said that the goal is to practice what Chauncey Starr once called "defensive engineering." Or perhaps more fundamental is the need for "preventive technology," which parallels what Ruth Davis says about preventive medicine:

> The solution to the problem is not in successive cures to successive science-caused problems; it is in their prevention. Unfortunately, cures for scientific ills are generally more interesting to scientists than is the prevention of those ills. We have the unhappy history of the medical community to show us the difficulties associated with trying to establish preventive medicine as a specialty.[11]

No amount of disciplined and imaginative foresight, however, can serve to anticipate all dangers. Because engineering projects are inherently experimental in nature, it is crucial for them to be monitored on an ongoing basis from the time they are put into effect. While individual practitioners cannot privately conduct full-blown environmental and social impact studies, they can choose to make the extra effort needed to keep in touch with the course of a project after it has officially left their hands. This is a mark of personal identification with one's work, a notion that leads to the next aspect of moral responsibility.

Moral Autonomy

People are morally autonomous when their moral conduct and principles of action are their own, in a special sense derived from Kant: Moral beliefs and attitudes should be held on the basis of critical reflection rather than passive adoption of the particular conventions of one's society, church, or profession. This is often what is meant by "authenticity" in one's commitment to moral values.

Those beliefs and attitudes, moreover, must be integrated into the core of an individual's personality in a manner that leads to committed action. They cannot be agreed to abstractly and formally, and adhered to merely verbally. Thus, just as one's principles are not passively absorbed from others when one is morally autonomous, so too one's actions are not treated as something alien and apart from oneself.

It is a comfortable illusion to think that in working for an employer, and thereby performing acts directly serving a company's interests, one is no longer morally and personally identified with one's actions. Selling one's labor and skills may make it

---

[11] Ruth M. Davis, "Preventative Technology: A Cure for Specific Ills," *Science* 188 (April 1975), p. 213.

seem that one has thereby disowned and forfeited power over one's actions.[12]

Viewing engineering as social experimentation can help one overcome this tendency and can help restore a sense of autonomous participation in one's work. As an experimenter, an engineer is exercising the sophisticated training that forms the core of his or her identity as a professional. Moreover, viewing an engineering project as an experiment that can result in unknown consequences should help inspire a critical and questioning attitude about the adequacy of current economic and safety standards. This also can lead to a greater sense of personal involvement with one's work.

The attitude of management plays a decisive role in how much moral autonomy engineers feel they have. It would be in the long-term interest of a high-technology firm to grant its engineers a great deal of latitude in exercising their professional judgment on moral issues relevant to their jobs (and, indeed, on technical issues as well). But the yardsticks by which a manager's performance is judged on a quarterly or yearly basis often discourage this. This is particularly true in our age of conglomerates, when near-term profitability is more important than consistent quality and long-term retention of satisfied customers.

In government-sponsored projects, it is frequently a deadline that becomes the ruling factor, along with fears of interagency or foreign competition. Tight schedules contributed to the loss of the space shuttle *Challenger,* as we shall see later.

Accordingly, engineers are compelled to look to their professional societies and other outside organizations for moral support. Yet it is no exaggeration to claim that the blue-collar worker with union backing has greater leverage at present in exercising moral autonomy than do many employed professionals. Professional societies, originally organized as learned societies dedicated to the exchange of technical information, lack comparable power to protect their members, although most engineers have no other group on which to rely for such protection. Only now is the need for moral and legal support of members in the exercise of their professional obligations being recognized by those societies.[13]

---

[12] John Lachs, "'I Only Work Here': Mediation and Irresponsibility," in *Ethics, Free Enterprise, and Public Policy,* ed. Richard T. DeGeorge and Joseph A. Pichler (New York: Oxford, 1978), pp. 201–13. Also see Elizabeth Wolgast, *Ethics of an Artificial Person: Lost Responsibility in Professions and Organizations* (Stanford: Stanford University Press, 1992).

[13] Stephen H. Unger, *Controlling Technology: Ethics and the Responsible Engineer,* 2nd ed. (New York: John Wiley & Sons, 1994).

Accountability

Finally, responsible people accept moral responsibility for their actions. Too often, "accountable" is understood in the overly narrow sense of being culpable and blameworthy for misdeeds. But the term more properly refers to the general disposition of being willing to submit one's actions to moral scrutiny and be open and responsive to the assessments of others. It involves a willingness to present morally cogent reasons for one's conduct when called upon to do so in appropriate circumstances.

Submission to an employer's authority, or any authority for that matter, creates in many people a narrowed sense of accountability for the consequences of their actions. This was documented by some famous experiments conducted by Stanley Milgram during the 1960s.[14] Subjects would come to a laboratory believing they were to participate in a memory and learning test. In one variation, two other people were involved, the "experimenter" and the "learner." The experimenter was regarded by the subject as an authority figure, representing the scientific community. He or she would give the subject orders to administer electric shocks to the "learner" whenever the latter failed in the memory test. The subject was told the shocks were to be increased in magnitude with each memory failure. All this, however, was a deception—a setup. There were no real shocks and the apparent "learner" and the "experimenter" were merely acting parts in a ruse designed to see how far the unknowing experimental subject was willing to go in following orders from an authority figure.

The results were astounding. When the subjects were placed in an adjoining room separated from the "learner" by a shaded glass window, more than half were willing to follow orders to the full extent: giving the maximum electric jolt of 450 volts. This was in spite of seeing the "learner," who was strapped in a chair, writhing in (apparent) agony. The same results occurred when the subjects were allowed to hear the (apparently) pained screams and protests of the "learner," screams and protests that became intense from 130 volts on. There was a striking difference, however, when subjects were placed in the same room within touching distance of the "learner." Then the number of subjects willing to continue to the maximum shock dropped by one-half.

---

[14] Stanley Milgram, *Obedience to Authority* (New York: Harper & Row, 1974).

Milgram explained these results by citing a strong psychological tendency in people to be willing to abandon personal accountability when placed under authority. He saw his subjects ascribing all initiative, and thereby all accountability, to what they viewed as legitimate authority. And he noted that the closer the physical proximity, the more difficult it becomes to divest oneself of personal accountability.

The divorce between causal influence and moral accountability is common in business and the professions, and engineering is no exception. Such a psychological schism is encouraged by several prominent features of contemporary engineering practice.

First, large-scale engineering projects involve fragmentation of work. Each person makes only a small contribution to something much larger. Moreover, the final product is often physically removed from one's immediate workplace, creating the kind of "distancing" that Milgram identified as encouraging a lessened sense of personal accountability.

Second, corresponding to the fragmentation of work is a vast diffusion of accountability within large institutions. The often massive bureaucracies within which so many engineers work are bound to diffuse and delimit areas of personal accountability within hierarchies of authority.

Third, there is frequently pressure to move on to a new project before the current one has been operating long enough to be observed carefully. This promotes a sense of being accountable only for meeting schedules.

Fourth, the contagion of malpractice suits currently afflicting the medical profession is carrying over into engineering. With this comes a crippling preoccupation with legalities, a preoccupation that makes one wary of becoming morally involved in matters beyond one's strictly defined institutional role.

We do not mean to underestimate the very real difficulties these conditions pose for engineers who seek to act as morally accountable people on their jobs. Much less do we wish to say engineers are blameworthy for all the harmful side effects of the projects on which they work, even though they partially cause those effects simply by working on the projects. That would be to confuse accountability with *blameworthiness,* and also to confuse *causal* responsibility with *moral* responsibility. But we do claim that engineers who endorse the perspective of engineering as a social experiment will find it more difficult to divorce themselves psychologically from personal responsibility for their work. Such an attitude will deepen their awareness of how engineers daily cooperate in a risky enterprise in which they exercise their personal expertise toward goals they are especially qualified to attain, and for which they are also accountable.

Babylon's Building Code, 1758 B.C.

Hammurabi, as king of Babylon, was concerned with strict order in his realm, and he decided that the builders of his time should also be governed by his laws. Thus he decreed:

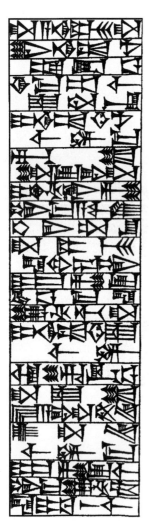

> If a builder has built a house for a man and has not made his work sound, and the house which he has built has fallen down and so caused the death of the householder, that builder shall be put to death. If it causes the death of the householder's son, they shall put that builder's son to death. If it causes the death of the householder's slave, he shall give slave for slave to the householder. If it destroys property he shall replace anything it has destroyed; and because he has not made sound the house which he has built and it has fallen down, he shall rebuild the house which has fallen down from his own property. If a builder has built a house for a man and does not make his work perfect and the wall bulges, that builder shall put that wall into sound condition at his own cost.[15]

A Balanced Outlook on Law

What should be the role of law in engineering, as viewed within our model of social experimentation? The legal regulations that apply to engineering and other professions are becoming more numerous and more specific all the time. We hear many complaints about this trend, and a major effort to deregulate various spheres of our lives is currently under way. Nevertheless, we continue to hear cries of "there ought to be a law" whenever a crisis occurs or a special interest is threatened.

This should not be surprising to us in the United States. We pride ourselves on being a nation that lives under the rule of law. We even delegate many of our decisions on ethical issues to an interpretation of laws. And yet this emphasis on law can cause problems in regard to ethical conduct beyond more practical issues usually cited by those who favor deregulation.

For example, one of the greatest moral problems in engineering, and one fostered by the very existence of minutely detailed rules, is that of *minimal compliance*. This can find its expression when companies or individuals search for loopholes in the law that will allow them to barely keep to its letter even while violating its spirit. Or hard-pressed engineers find it convenient to refer to standards with ready-made specifications as a substitute for original thought, perpetuating the "handbook mentality" and the repetition of mistakes. Minimal compliance led to the tragedy

---

[15] Hammurabi, *The Code of Hammurabi,* trans. R. F. Harper (Chicago: University of Chicago Press, 1904).

of the *Titanic*: Why should that ship have been equipped with enough lifeboats to accommodate all its passengers and crew when British regulations at the time required only a lower minimum, albeit with smaller ships in mind?

On the other hand, remedying the situation by continually updating laws or regulations with further specifications may also be counterproductive. Not only will the law inevitably lag behind changes in technology and produce a judicial vacuum, there is also the danger of overburdening the rules and the regulators. As Robert Kates puts it:

> If cooperation is not forthcoming—if the manufacturer, for example, falsifies or fails to conduct safety tests—there is something akin to the law of infinite regress in which the regulator must intrude more and more expensively into the data collection and evaluation process. In the end, the magnitude of the task overwhelms the regulators.[16]

Lawmakers cannot be expected always to keep up with technological development. Nor would we necessarily want to see laws changed upon each innovation. Instead we empower rule-making and inspection agencies—the Food and Drug Administration (FDA), the Federal Aviation Agency (FAA), and the Environmental Protection Agency (EPA) are examples of these in the United States—to fill the void. Though they are independent and belong to neither the judicial nor the executive branches of government, their rules have, for all practical purposes, the effect of law.

Industry tends to complain that excessive restrictions are imposed on it by regulatory agencies. But one needs to reflect on why regulations may have been necessary in the first place. Take, for example, the U.S. Consumer Product Safety Commission's rule for baby cribs, which specifies that "the distance between components (such as slats, spindles, crib rods, and corner posts) shall not be greater than $2^3/_8$ inches at any point." This rule came about because some manufacturers of baby furniture had neglected to consider the danger of babies strangling in cribs or had neglected to measure the size of babies' heads.[17]

Again, why must regulations be so specific when broad statements would appear to make more sense? When the EPA adopted rules for asbestos emissions in 1971, it was recognized that strict numerical standards would be impossible to promulgate.

---

[16] Robert W. Kates, ed., *Managing Technological Hazards: Research Needs and Opportunities* (Boulder, CO: Institute of Behavioral Science, University of Colorado, 1977), p. 32.

[17] William W. Lowrance, *Of Acceptable Risk* (Los Altos, CA: William Kaufmann, 1976), p. 134.

Asbestos dispersal and intake, for example, are difficult to measure in the field. So, being reasonable, the EPA specified a set of work practices to keep emissions to a minimum—that asbestos should be wetted down before handling, for example, and disposed of carefully. The building industry called for more specifics. Modifications in the Clean Air Act eventually permitted EPA to issue enforceable rules on work practices, and later the Occupational Safety and Health Administration also became involved.

Society's attempts at regulation have indeed often failed, but it would be wrong to write off rule-making and rule-following as futile. Good laws, effectively enforced, clearly produce benefits. They authoritatively establish reasonable minimal standards of professional conduct and provide at least a self-interested motive for most people and corporations to comply. Moreover, they serve as a powerful support and defense for those who wish to act ethically in situations where ethical conduct might be less than welcome. By being able to point to a pertinent law, one can feel more free to act as a responsible engineer.

Engineering as social experimentation can provide engineers with a proper perspective on laws and regulations in that rules that govern engineering practice should not be devised or construed as rules of a game but as rules of responsible experimentation. Such a view places proper responsibility on the engineer who is intimately connected with his or her "experiment" and responsible for its safe conduct; moreover, it suggests the following conclusions: For safeguarding the public, precise rules and enforceable sanctions are appropriate components of well-established and regularly reexamined engineering procedures. Little of an experimental nature is probably occurring in such standard activities, and the type of professional conduct required is most likely very clear. In areas where experimentation is involved more substantially, however, rules must not attempt to cover all possible outcomes of an experiment, nor must they force engineers to adopt rigidly specified courses of action. It is here that regulations should be broad, but written to hold engineers accountable for their decisions. Through their professional societies, engineers should also play an active role in establishing (or changing) enforceable rules as well as in enforcing them, taking great care to forestall conflicts of interest (see Discussion Topic 5, on the Hydrolevel case).

Industrial Standards

There is one area in which industry usually welcomes greater specificity, and that is in regard to standards. Standards facilitate the interchange of components, they serve as ready-made substitutes for lengthy design specifications, and they decrease production costs.

Table 3–1 Types of Standards

Criterion	Purpose	Selected examples
Uniformity of physical properties and functions	Accuracy in measurement, interchangeability, ease of handling	Standards of weights, screw dimensions, standard time, film size
Safety and reliability	Prevention of injury, death, and loss of income or property	National Electric Code, boiler code, methods of handling toxic wastes
Quality of product	Fair value for price	Plywood grades, lamp life
Quality of personnel and service	Competence in carrying out tasks	Accreditation of schools, professional licenses
Use of accepted procedures	Sound design, ease of communications	Drawing symbols, test proecdures
Separability	Freedom from interference	Highway lane markings, radio frequency bands
Quality procedures approved by ISO, the International Standards Organization	Assurance of product acceptance in member countries	Quality of products, work, certificates, and degrees

A selection of modular telephone adaptors offered by Magellan's:

© Magellan's. A small section from 35 types shown in Magellan's 1999 catalogs. Reprinted with permission.

"Wouldn't it be nice if they could agree on a common telephone jack?"

Standards consist of explicit specifications that, when followed with care, ensure that stated criteria for interchangeability and quality will be attained. Examples range from automobile tire sizes and load ratings, to computer languages. Table 3–1 lists purposes of standards and gives some examples to illustrate those purposes.

Standards are established by companies for in-house use and by professional associations and trade associations for industry-wide use. They may also be prescribed as parts of laws and official regulations, for example, as in mandatory standards, which become necessary upon lack of adherence to voluntary standards.

Standards help not only the manufacturers, they also benefit the client and the public. They preserve some competitiveness in industry by reducing overemphasis on name brands and giving the smaller manufacturer a chance to compete. They ensure a measure of quality and thus facilitate more realistic trade-off decisions. International standards are becoming a necessity in European and world trade. An interesting approach has been adopted by the International Standards Organization (ISO) that replaces the detailed national specifications for a plethora of products with statements of procedures a manufacturer guarantees to carry out in assuring quality products.

Standards have been a hindrance at times. For many years, they were mostly descriptive, specifying, for instance, how many joists of what size should support a given type of floor. Clearly such standards tended to stifle innovation. The move to performance standards, which in the case of a floor may specify only the required load-bearing capacity, has alleviated that problem somewhat. But other difficulties can arise when special interests (for example, manufacturers, trade unions, exporters and importers) manage to impose unnecessary provisions on standards, or remove important provisions from them, to secure their own narrow self-interest. Requiring metal conduits for home wiring is one example of this problem. Modern conductor coverings have eliminated the need for metal conduit in numerous applications, but many localities still require it. Its use sells more conduit and employs more electricians for installation.

There are standards nowadays for practically everything, it seems, and consequently we often assume that stricter regulation exists than may actually be the case. The public tends to trust the National Electrical Code in all matters of power distribution and wiring, but how many people realize that this code, issued by the National Fire Protection Association, is primarily oriented toward fire hazards? Only recently have its provisions against electric shock begun to be strengthened. Few consumers know that an Underwriter Laboratories seal prominently affixed to the cord of an electrical appliance may pertain only to the cord and not to the rest of the device. In a similar vein, a patent notation inscribed on the handle of a product may refer just to the handle, and then possibly only to the design of the handle's appearance.

Sometimes standards are thought to apply when in actuality there is no standard at all. An example can be found in the widely varying worth and quality of academic degrees—doctorates are even available from mail-order houses. Product appearances can be misleading in this respect. Years ago, when competing foreign firms were attempting to corner the South American market for electrical fixtures and appliances, one manufacturing company had a shrewd idea. It equipped its light bulbs with extra-long bases and threads. These would fit into the competitors' shorter lamp sockets and its own deep sockets. But the competitors' bulbs would not fit into the deeper sockets of its own fixtures (see Figure 3–1). Yet so far as the unsuspecting consumer was concerned, all the light bulbs and sockets continued to look alike.

During the introduction of novel products, there is frequently a period during which the consumer is at a disadvantage, not knowing which type or size of magnetic recording tape, word pro-

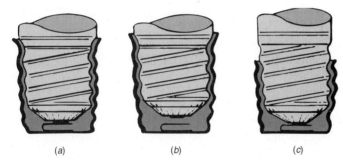

Figure 3–1
The light bulb story. (*a*) Long base, deep socket: firm contact. (*b*) Short base, deep socket: no contact. (*c*) Long base, shallow socket: firm contact.

cessing program, or camera lens mount will win in the long run and thus make a recently purchased product prematurely obsolete or nonrepairable. Sometimes a particular design stays in the front long enough until it becomes the standard, as happened to Hayes modems and their command structure. More recently though, a joint effort by otherwise competing photographic film manufacturers resulted in an agreed-upon standard before they introduced the Advanced Photo System (APS) with new standards for film cassettes and processing.

Discussion Topics

1. A common excuse for carrying out a morally questionable project is, "If I don't do it, somebody else will." This rationale may be tempting for engineers who typically work in situations where someone else might be ready to replace them on a project. Do you view it as a legitimate excuse for engaging in projects that might be unethical? (In your answer, comment on the concept of responsible conduct developed in this section.)

2. Another commonly used phrase, "I only work here," implies that one is not personally accountable for the company rules since one does not make them. It also suggests that one wishes to restrict one's area of responsibility within tight bounds as defined by those rules. In light of the discussion in this section, respond to the potential implications of this phrase and the attitude represented by it when exhibited by engineers.

3. Threats to a sense of personal responsibility are neither unique to nor more acute for engineers than they are for others involved with engineering and its results. The reason is that, in general, public accountability also tends to lessen as professional roles become narrowly differentiated. With this in mind, critique each

of the remarks made in the following dialog. Is the remark true, or partially true? What needs to be added to make it accurate?

ENGINEER: My responsibility is to receive directives and to create products within specifications set by others. The decision about what products to make and their general specifications are economic in nature and made by management.

SCIENTIST: My responsibility is to gain knowledge. How the knowledge is applied is an economic decision made by management, or else a political decision made by elected representatives in government.

MANAGER: My responsibility is solely to make profits for stockholders.

STOCKHOLDER: I invest my money for the purpose of making a profit. It is up to managers to make decisions about the directions of technological development.

CONSUMER: My responsibility is to my family. Government should make sure corporations do not harm me with dangerous products, harmful side effects of technology, or dishonest claims.

GOVERNMENT REGULATOR: By current reckoning, government has strangled the economy through overregulation of business. Accordingly, at present on my job, especially given decreasing budget allotments, I must back off from the idea that business should be policed and urge corporations to assume greater public responsibility.

4. In 1975, Hydrolevel Corporation brought suit against the American Society of Mechanical Engineers (ASME), charging that two ASME volunteers, acting as agents of ASME, had conspired to interpret a section of ASME's Boiler and Pressure Vessel Code in such a manner that Hydrolevel's low-water fuel cutoff for boilers could not compete with the devices built by the employers of the two volunteers. On May 17, 1982, the Supreme Court upheld the lower courts that had found ASME guilty of violating antitrust provisions and had opened the way for awarding treble damages. (A U.S. district court's award of $7.5 million had been found excessive by the court of appeals and later the award was set at $4.74 million, including legal costs.)

   Writing on behalf of the six-to-three majority, Justice Harry A. Blackmun said:

   When ASME's agents act in its name, they are able to affect the lives of large numbers of people and the competitive fortunes of businesses throughout the country. By holding ASME liable under the antitrust laws for the antitrust violations of its agents committed with apparent authority, we recognize the important role of ASME and its agents in the economy, and we help to ensure that standard-setting

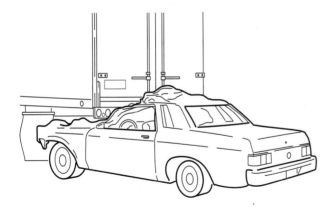

organizations will act with care when they permit their agents to speak for them.

Acquaint yourself with the particulars of this case and discuss it as an illustration of the possible misuses of standards.[18]

5. Mismatched bumpers: Ought there to be a law? What happens when a passenger car rear-ends a truck or a sports utility vehicle (SUV)? The bumpers usually ride at different heights, so even modest collisions can result in major repair bills. (At high speed, with the front of the car nose down upon braking, people in convertibles have been decapitated upon contact devoid of protection by bumpers.) The men and women at Volvo recognized the problem long ago—we have observed that their trucks usually have low bumpers front *and* rear. Discuss how other companies building and selling trucks and high-riding vehicles can be induced to follow Volvo's example. Should older vehicles be retrofitted with lower bumpers or guards? Must we have yet another law because otherwise industry will not act on its own?

6. If someone were to ask you to describe codes of ethics for engineers in terms of the three columns provided in Table 3–1, what entries would you provide?

The *Challenger* Case

Several months before the destruction of *Challenger*, NASA historian Alex Roland wrote the following in a critical piece about the space shuttle program:

18 Stephen H. Unger, *Controlling Technology: Ethics and the Responsible Engineer*, 2nd ed. (New York: John Wiley & Sons, 1994), pp. 210–15. See also Paula Wells, Hardy Jones, and Michael Davis, *Conflicts of Interest in Engineering* (Dubuque, IA: Kendall/Hunt, 1986).

The American taxpayer bet about $14 billion on the shuttle. NASA bet its reputation. The Air Force bet its reconnaissance capability. The astronauts bet their lives. We all took a chance.

When John Young and Robert Crippen climbed aboard the orbiter *Columbia* on April 12, 1981, for the first shuttle launch, they took a bigger chance than any astronaut before them. Never had Americans been asked to go on a launch vehicle's maiden voyage. Never had astronauts ridden solid propellant rockets. Never had Americans depended on an engine untested in flight.[19]

Most of Alex Roland's criticism was directed at the economic and political aspects of the shuttle program that was supposed to become a self-supporting operation but never gave any indication of being able to reach that goal. Without a national consensus to back it, the program became a victim of year-by-year funding politics.

The *Columbia* and its sister ships, *Challenger, Discovery,* and *Endeavor,* are delta-wing craft with a huge payload bay. Early, sleek designs had to be abandoned to satisfy U.S. Air Force requirements when the Air Force was ordered to use the NASA shuttle instead of its own expendable rockets for launching satellites and other missions. As shown in Figure 3–2, each orbiter has three main engines fueled by several million pounds of liquid hydrogen; the fuel is carried in an immense, external, divided fuel tank, which is jettisoned when empty. During liftoff, the main engines fire for about 8.5 minutes, although during the first 2 minutes of the launch, much of the thrust is provided by two booster rockets. These are of the solid-fuel type, each burning a one-million-pound load of a mixture of aluminum, potassium chloride, and iron oxide.

The casing of each booster rocket is about 150 feet long and 12 feet in diameter. It consists of cylindrical segments that are assembled at the launch site. The four field joints use seals composed of pairs of O-rings made of vulcanized rubber. The O-rings work in conjunction with a putty barrier of zinc chromide.

The shuttle flights were successful, though not as frequent as had been hoped. NASA tried hard to portray the shuttle program as an operational system that could pay for itself. Some Reagan administration officials had even suggested that the operations be turned over to an airline. Aerospace engineers intimately involved in designing, manufacturing, assembling, testing, and operating the shuttle still regarded it as an experimental undertaking in 1986. These engineers were employees of manufacturers, such as Rockwell International (orbiter and main rocket)

---

[19] Alex Roland, "The Shuttle, Triumph or Turkey?" *Discover,* November 1985, pp. 29–49.

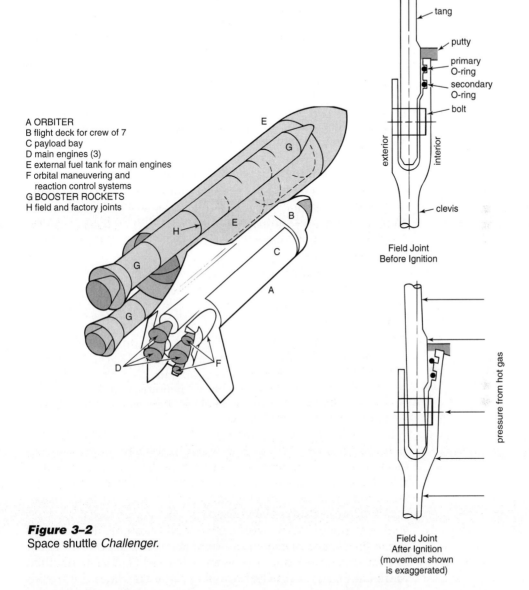

A ORBITER
B flight deck for crew of 7
C payload bay
D main engines (3)
E external fuel tank for main engines
F orbital maneuvering and
  reaction control systems
G BOOSTER ROCKETS
H field and factory joints

tang
putty
primary
O-ring
secondary
O-ring
bolt
clevis
exterior
interior

Field Joint
Before Ignition

pressure from hot gas

Field Joint
After Ignition
(movement shown
is exaggerated)

Figure 3–2
Space shuttle *Challenger*.

and Morton-Thiokol (booster rockets), or they worked for NASA
at one of its several centers: Marshall Space Flight Center,
Huntsville, Alabama (responsible for the propulsion system);
Kennedy Space Center, Cape Kennedy, Florida (launch operations); Johnson Space Center, Houston, Texas (flight control);
and the office of the Chief Engineer, Washington, D.C. (overall
responsibility for safety, among other duties).

After embarrassing delays, *Challenger*'s first flight for 1986 was set for Tuesday morning, January 28. But Allan J. McDonald, who represented Morton-Thiokol at Cape Kennedy, was worried about the freezing temperatures predicted for the night. As his company's director of the solid-rocket booster project, he knew of difficulties that had been experienced with the field joints on a previous cold-weather launch when the temperature had been mild compared to what was forecast. He therefore arranged a teleconference so that NASA engineers could confer with Morton-Thiokol engineers at their plant in Utah.

Arnold Thompson and Roger Boisjoly, two seal experts at Morton-Thiokol, explained to their own colleagues and managers as well as the NASA representatives how, upon launch, the booster rocket walls bulge and the combustion gases can blow past one or even both of the O-rings that make up the field joints (see Figure 3–2). The rings char and erode, as had been observed on many previous flights. In cold weather, the problem is aggravated because the rings and the putty packing are less pliable then. But

> . . . only limited consideration was given to the past history of O-ring damage in terms of temperature. The managers compared as a function of temperature the flights for which thermal distress of O-rings had been observed [Figure 3–3]—not the frequency of occurrence based on all flights. When the entire history of flight experience is considered, including "normal" flights with no erosion or blow-by, the comparison is substantially different [Figure 3–4]. Consideration of the entire launch temperature history indicates that the probability of O-ring distress is increased to almost a certainty if the temperature of the joint is less than 65.[20]

If the graph depicted in Figure 3–4 and other supporting renditions (such as Figure 3–5 of later date) had been available at the meeting, upper management might just have seen the problem of cold O-rings more clearly, but with launch time approaching, deliberations were cut short.

The engineering managers, Bob Lund (vice president of engineering) and Joe Kilminster (vice president for booster rockets), agreed that there was a problem with safety. The team from Marshall Space Flight Center was incredulous. Since the specifications called for an operating temperature of the solid fuel prior to combustion of 40 to 90 degrees Fahrenheit, one could surely

---

[20] Rogers Commission Report, *Report of the Presidential Commission on the Space Shuttle* Challenger *Accident* (Washington, DC: U.S. Government Printing Office, 1986).

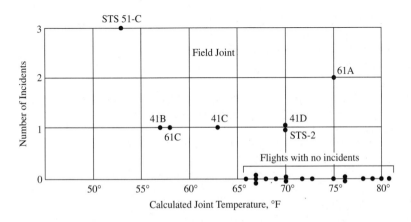

Figure 3–3
Plot of flights *with* incidents of O-ring thermal distress: Number of O-rings affected plotted against temperature. (*Source: Rogers Commission Report,* Report of the Presidential Commission on the Space Shuttle *Challenger* Accident [*Washington, DC: U.S. Government Printing Office, 1986*].)

Figure 3–4
Plot of flights *with and without* incidents of O-ring thermal distress. (*Source: Rogers Commission Report,* Report of the Presidential Commission on the Space Shuttle *Challenger* Accident [*Washington, DC: U.S. Government Printing Office, 1986*].)

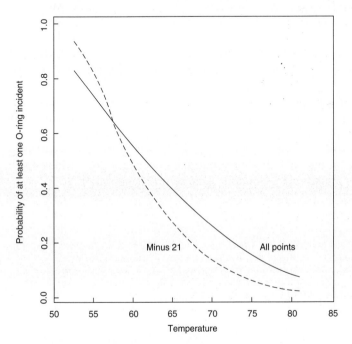

Figure 3–5
Binomial-logit model based on all data remaining after removing flight #61A in Figure 3–4, and also *with* flight 61A (data point #21). The ordinate gives the probability of distress of at least one O-ring. (*Source: Siddharta R. Dalal, Edward B. Fowlkes, and Bruce Hoadley, "Risk Analysis of the Space Shuttle: Pre-Challenger Prediction of Failure,"* Journal of the American Statistical Association *84,* [*December 1989*], pp. 945–57.)

allow lower or higher outdoor temperatures, notwithstanding Boisjoly's testimony and recommendation that no launch should occur at less than 53 degrees. They were clearly annoyed at facing yet another postponement.

Top executives of Morton-Thiokol were also sitting in on the teleconference. Their concern was the image of the company, which was in the process of negotiating a renewal of the booster rocket contract with NASA. During a recess, Senior Vice President Jerry Mason turned to Bob Lund and told him "to take off your engineering hat and put on your management hat." It was a subsequent vote (of the managers only) that produced the company's official finding that the seals could not be shown to be unsafe. The engineers' judgment was not considered sufficiently weighty. At Cape Kennedy, Allan McDonald refused to sign the formal recommendation to launch; Joe Kilminster had to. Accounts of the *Challenger* disaster tell of the cold Tuesday morning, the high seas that forced the recovery ships to seek coastal shelter, the ice at the launch site, and the concern expressed by Rockwell engineers that the ice might shatter and hit the orbiter or rocket casings.[21] The inability of these engineers to *prove* that the liftoff would be unsafe was taken by NASA as an approval by Rockwell to launch.

The countdown ended at 11:38 A.M. The temperature had risen to 36 degrees. As the rockets carrying *Challenger* rose from the ground, cameras recorded puffs of smoke that emanated from one of the field joints on the right booster rocket. Soon these turned into a flame that hit the external fuel tank and a strut holding the booster rocket. The hydrogen in the tank caught fire; the booster rocket broke loose and smashed into *Challenger*'s wing and then into the external fuel tank. At 76 seconds into the flight, by the time *Challenger* and its rockets had reached 50,000 feet, it was totally engulfed in a fireball. The crew cabin separated and fell into the ocean, killing all aboard: mission commander Francis (Dick) Scobee; pilot Michael Smith; mission specialists Gregory Jarvis, Ronald McNair, Ellison Onizuka, and Judith Resnick; and "teacher in space" Christa MacAuliffe.

President Reagan was to give his State of the Union message later that day. He had to change the tone of his prepared remarks on the shuttle flight and its first civilian passenger.

---

[21] Malcolm McConnell, *Challenger, a Major Malfunction* (Garden City, NY: Doubleday, 1987); Rosa Lynn B. Pinkus, Larry J. Shuman, Norman P. Hummon, and Harvey Wolfe, *Engineering Ethics: Balancing Cost, Schedule, and Risk—Lessons Learned from the Space Shuttle* (Cambridge: Cambridge University Press, 1997).

Safety Issues

Unlike the *three-stage rockets* that carried astronauts to the moon, the *space shuttle* could be involved in a simultaneous (inadvertent) ignition of all fuel carried aloft. An explosion close to the ground can have catastrophic effects. The crew had no escape mechanism, although McDonnell Douglas, in a losing shuttle proposal, had designed an abort module with its own thruster. It would have allowed the separation of the orbiter, triggered (among other events) by a field-joint leak. But such a safety measure was rejected as too expensive because of an accompanying reduction in payload.

Working with such constraints, why was safe operation not stressed more? First of all, we must remember that the shuttle program was indeed still a truly experimental and research undertaking. Next, it is quite clear that the members of the crews knew that they were embarking on dangerous missions. But it has also been revealed that the *Challenger* astronauts were not informed of particular problems such as the field joints. They were not asked for their consent to be launched under circumstances that experienced engineers had claimed to be unsafe.

The reason for the rather cavalier attitude toward safety is revealed in the way NASA assessed the system's reliability. For instance, recovered booster rocket casings had indicated that the field-joint seals had been damaged in many of the earlier flights. The waivers necessary to proceed with launches had become mere gestures. Richard Feynman made the following observations as a member of the Presidential Commission on the Space Shuttle *Challenger* Accident (called the Rogers Commission, after its chairman):

> I read all of these [NASA flight readiness] reviews and they agonize whether they can go even though they had some blow-by in the seal or they had a cracked blade in the pump of one of the engines . . . and they decide "yes." Then it flies and nothing happens. Then it is suggested . . . that the risk is no longer so high. For the next flight we can lower our standards a little bit because we got away with it last time. . . . It is a kind of Russian roulette.[22]

Since the early days of unmanned space flight, about 1 in every 25 solid-fuel rocket boosters had failed. Given improvements over the years, Feynman thought that 1 in every 50 to 100 might be a reasonable estimate now. Yet NASA counted on only

---

[22] Rogers Commission Report, *Report of the Presidential Commission.*

1 crash in every 100,000 launches. Queried about these figures, NASA Chief Engineer Milton Silveira answered: "We don't use that number as a management tool. We know that the probability of failure is always sitting there."[23] So where was this number used? In a risk analysis needed by the Department of Energy to assure everyone that it would be safe to use small atomic reactors as power sources on deep-space probes and to carry both aloft on a space shuttle. As luck would have it, *Challenger* was not to carry the 47.6 pounds of lethal plutonium-238 until its next mission with the Galileo probe on board.[24]

Another area of concern was NASA's unwillingness to wait out risky weather. When serving as weather observer, astronaut John Young was dismayed to find his recommendations to postpone launches disregarded several times. Things had not changed much by March 26, 1987, when NASA ignored its devices monitoring electric storm conditions, launched a Navy communications satellite atop an Atlas-Centaur rocket, and had to destroy the $160 million system when it veered off course after being hit by lightning. The storm monitors had been installed after an *Apollo* command module almost had its moon trip aborted 18 years earlier because of storm conditions. Weather, incidentally, could be held partially responsible for the shuttle disaster because a strong wind shear may have contributed to the rupturing of the weakened O-rings.[25]

Veteran astronauts were also dismayed at NASA management's decision to land at Cape Kennedy as often as possible despite its unfavorable landing conditions, including strong crosswinds and changeable weather. The alternative, Edwards Air Force Base in California, is a better landing place but requires a piggyback ride for the shuttle on a Boeing 747 home to Florida. This costs time and money.

In 1982, Albert Flores had conducted a study of safety concerns at the Johnson Space Center. He found its engineers to be strongly committed to safety in all aspects of design. When they were asked how managers might further improve safety awareness, there were few concrete suggestions but many comments

---

[23] Eliot Marshall, "Feynman Issues His Own Shuttle Report, Attacking NASA Risk Estimates," *Science* 232 (June 1986), p. 1596. See also Richard P. Feynman's rendition on the shuttle fiasco and the Rogers Commission workings in *What Do You Care What Other People Think?*, as told to Ralph Leighton (New York: W. W. Norton & Co., 1988).

[24] Karl Grossman, "Red Tape and Radioactivity," *Common Cause,* July–August 1986, pp. 24–27.

[25] Trudy E. Bell, "Wind Shear Cited as Likely Factor in Shuttle Disaster," *The Institute* (IEEE) 11 (May 1987), p. 1. For effects of lightning, see Eliot Marshall, "Lightning Strikes Twice at NASA," *Science* 236 (May 1987), p. 903.

on how safety concerns were ignored or negatively impacted by management. One engineer was quoted as saying, "A small amount of professional safety effort and upper management support can cause a quantum safety improvement with little expense."[26] This points to the important role of management in building a strong sense of responsibility for safety first and for schedules second.

The space shuttle's field joints are designated "criticality 1," which means there is no backup. Therefore, a leaky field joint will result in failure of the mission and loss of life. There are 700 items of criticality 1 on the shuttle. A problem with any one of them should have been cause enough to do more than just launch more shuttles without modification while working on a better system. Improved seal designs had already been developed, but the new rockets would not have been ready for some time. In the meantime, the old booster rockets should have been recalled.

At Morton-Thiokol, Roger Boisjoly's personal concern had been heightened by his memory of the DC-10 crash over Paris. That accident had shown him how known defects can be disregarded in a complex organization. For this reason, he had started a journal in which he recorded all events associated with the seals.[27] But he probably did not feel that he had the kind of professional backing that would allow him to go beyond his organization directly to the astronauts.

In several respects, the ethical issues in the *Challenger* case resemble those of other such cases. Concern for safety gave way to institutional posturing. Danger signals did not go beyond Morton-Thiokol and Marshall Space Flight Center in the *Challenger* case. No effective recall was instituted. There were concerned engineers who spoke out, but ultimately they felt it only proper to submit to management decisions.

One notable aspect of the *Challenger* case is the late-hour teleconference that Allan McDonald had arranged from the *Challenger* launch site to get knowledgeable engineers to discuss the seal problem from a technical viewpoint. This tense conference did not involve lengthy discussions of ethics, but it revealed the virtues (or lack thereof) that allow us to distinguish between the "right stuff" and the "wrong stuff." This is well described by one aerospace engineer as arrogance, specifically "The arrogance that prompts higher-level decision makers to pretend that factors other than engineering judgment should influence flight safety

---

[26] Albert Flores, ed., *Designing for Safety: Engineering Ethics in Organizational Contexts* (Troy, NY: Rensselaer Polytechnic Institute, 1982), p. 79; Feynman, *What Do You Care What Other People Think?*

[27] Caroline Whitbeck, "Moral Responsibility and the Working Engineer," *Books and Religion* 15 (March–April 1987), p. 3.

decisions and, more important, the arrogance that rationalizes overruling the engineering judgment of engineers close to the problem by those whose expertise is naive and superficial by comparison."[28] Included, surely, is the arrogance of those who reversed NASA's (paraphrased) motto, "Don't fly if it cannot be shown to be safe" to "Fly unless it can be shown not to be safe."

At Morton-Thiokol, some of the vice presidents in the space division have been demoted. The engineers who were outspoken at the prelaunch teleconference and again before the Rogers Commission kept their jobs at the company because of congressional pressure, but their jobs are of a pro forma nature. In a speech to engineering students at the Massachusetts Institute of Technology a year after the *Challenger* disaster, Roger Boisjoly said: "I have been asked by some if I would testify again if I knew in advance of the potential consequences to me and my career. My answer is always an immediate yes. I couldn't live with any self-respect if I tailored my actions based upon potential personal consequences as a result of my honorable actions."[29]

Today NASA has a policy that allows aerospace workers with concerns to report them anonymously to the Batelle Memorial Institute in Columbus, Ohio, but open disagreement still invited harassment for a number of years.

Discussion Topics

1. Chairman Rogers asked Bob Lund: "Why did you change your decision [that the seals would not hold up] when you changed hats?" What might motivate you, as a midlevel manager, to go along with top management when told to "take off your engineering hat and put on your management hat"? Applying the engineering-as-experimentation model, what might responsible experimenters have done in response to the question?

2. Under what conditions would you say it is safe to launch a shuttle without an escape mechanism for the crew?

3. Discuss the role of the astronauts in shuttle safety. To what extent should they (or at least the orbiter commanders) have involved themselves more actively in looking for safety defects in design or operation?

4. Consider the following actions or recommendations and suggest a plan of action to bring about safer designs and operations in a complex organization.

---

[28] Calvin E. Moeller, "*Challenger* Catastrophe," *Los Angeles Times,* Letters to the Editor, March 11, 1986.

[29] Roger M. Boisjoly, Speech on shuttle disaster delivered to MIT students, January 7, 1987. Printed in *Books and Religion* 15 (March–April 1987), p. 3.

(a) Lawrence Mulloy represented Marshall Space Flight Center at Cape Kennedy. He did not tell Arnold Aldrich from the National Space Transportation Program at Johnson Space Center about the discussions regarding the field-joint seals even though Aldrich had the responsibility of clearing *Challenger* for launch. Why? Because the seals were "a Level III issue," and Mulloy was at Level III, while Aldrich was at a higher level (Level II), which ought not to be bothered with such details.

(b) The Rogers Commission recommended that an independent safety organization directly responsible to the NASA administrator be established. An anonymous reporting scheme now exists for aerospace industry employees working on NASA projects.

(c) Tom Peters advises managers to "involve everyone in everything.... Boldly assert that there is no limit to what the average person can accomplish if thoroughly involved." [30]

5. On October 4, 1930, the British airship *R 101* crashed about eight hours into its maiden voyage to India. Of the 54 persons aboard, only 6 survived. Throughout the craft's design and construction, Air Ministry officials and their engineers had been driven by strong political and competitive forces. Nevil Shute, who had worked on the rival, commercial *R 100,* wrote in his memoir, *Slide Rule,* that "if just one of [the men at the Air Ministry] had stood up [at a conference with Lord Thomson] and had said, 'This thing won't work, and I'll be no party to it. I'm sorry, gentlemen, but if you do this, I'm resigning,' ... the disaster would almost certainly have been averted. It was not said, because the men in question put their jobs before their duty."[31] Examine the *R 101* case, and compare it with the *Challenger* case, including the pressures not to delay the flight.

---

[30] Tom Peters, *Thriving on Chaos* (New York: Alfred A. Knopf, 1987).

[31] Nevil Shute, *Slide Rule* (New York: William Morrow, 1954), p. 140. Also see Henry Cord Meyer, "Politics, Personality, and Technology: Airships in the Manipulations of Dr. Hugo Eckener and Lord Thomson, 1919–1930," *Aerospace Historian* (September 1981), 165–72; Arthur M. Squires, *The Tender Ship: Governmental Management of Technological Change* (Boston: Birkhauser, 1986).

4

Commitment to Safety

Pilot Dan Gellert was flying an Eastern Airlines Lockheed L-1011, cruising at an altitude of 10,000 feet, when he inadvertently dropped his flight plan.[1] Being on autopilot control, he casually leaned down to pick it up, and inadvertantly bumped the control stick. This should not have mattered, but immediately the plane went into a steep dive, terrifying the 230 passengers. Badly shaken himself, Gellert was nevertheless able to grab the control stick and ease the plane back onto course. Though much altitude had been lost, the altimeter still read 10,000 feet.

Not long before this incident, one of Gellert's colleagues had been in a flight trainer when the autopilot and the flight trainer disengaged, producing a crash on an automatic landing approach. Fortunately it all happened in simulation. But just a short time later, an Eastern Airlines L-1011 actually crashed on approach to Miami. On that flight, there seemed to have been some problem with the landing gear, so the plane had been placed on autopilot at 2000 feet while the crew investigated the trouble. Four minutes later, after apparently losing altitude without warning while the crew was distracted, it crashed in the Everglades, killing 103 people.

A year later, Gellert was again flying an L-1011 and the autopilot disengaged once more when it should not have done so. The plane was supposedly at 500 feet and on the proper glide slope to landing as it broke through a cloud cover. Suddenly realizing it was only at 200 feet and above a densely populated area, the crew had to engage the plane's full takeoff power to make the runway safely.

The L-1011 incidents point out how vulnerable our intricate machines and control systems can be, how they can malfunction because of unanticipated circumstances, and how important it is

---

[1] Dan Gellert, "Whistle-Blower: Dan Gellert, Airline Pilot," *The Civil Liberties Review,* September 1978.

to design for proper human–machine interactions whenever human safety is involved. In this chapter, we discuss the role of safety as seen by the public and the engineer.

Typically, several groups of people are involved in safety issues, each with its own interests at stake. If we now consider that within each group there are differences of opinion regarding what is safe and what is not, it becomes obvious that "safety" can be an elusive term, as can "risk." Following a look at these basic concepts, we will then turn to safety and risk assessment and methods of reducing risk. Finally, by way of examining the nuclear power plant accidents at Three Mile Island and Chernobyl, we will consider the implications of an ever-growing complexity in engineered systems and the ultimate need for safe exits.

Safety and Risk

We demand safe products and services because we do not wish to be threatened by potential harm, but we also realize that we may have to pay for this safety. To complicate matters, what may be safe enough for one person may not be for someone else—either because of different perceptions about what is safe or because of different predispositions to harm. A power saw in the hands of a child will never be as safe as it can be in the hands of an adult. And a sick adult is more prone to suffer ill effects from air pollution than is a healthy adult.

Absolute safety, in the senses of (*a*) entirely risk-free activities and products or (*b*) a degree of safety that satisfies all individuals or groups under all conditions, is neither attainable nor affordable. Yet it is important that we come to some understanding of what we mean by safety.

The Concept of Safety

One approach to defining "safety" would be to render the notion thoroughly subjective by defining it in terms of whatever risks a person judges to be acceptable. Such a definition was given by William W. Lowrance: "*A thing is safe if its risks are judged to be acceptable.*"[2] This approach helps underscore the notion that judgments about safety are tacitly value judgments about what is acceptable risk to a given person or group. Differences in appraisals of safety are thus correctly seen as reflecting differences in values.

Lowrance's definition, however, needs to be modified, for it departs too far from our common understanding of safety. This

---

[2] William W. Lowrance, *Of Acceptable Risk* (Los Altos, CA: William Kaufmann, 1976), p. 8.

can be shown if we consider three types of situations that can arise. Imagine, first, a case where we seriously underestimate the risks of something, say of using a toaster we see at a garage sale. On the basis of that mistaken view, we judge it to be very safe and buy it. On taking it home and trying to make toast with it, we end up in the hospital with a severe electric shock or burn. Using the ordinary notion of safety, we conclude we were wrong in our earlier judgment: The toaster was not safe at all! Given our values and our needs, its risks should not have been judged acceptable earlier. Yet by Lowrance's definition, we would be forced to say that prior to the accident the toaster was entirely safe since, after all, at that time we had judged the risks to be acceptable.

Consider, second, the case where we grossly overestimate the risks of something. For example, suppose that we irrationally think fluoride in drinking water will kill a third of the populace. According to Lowrance's definition, the fluoridated water is unsafe, since we judge its risks to be unacceptable. It would be impossible, moreover, for someone to reason with us to prove that the water is actually safe. For again, according to his definition, the water became unsafe the moment we judged the risks of using it to be unacceptable for us.

Third, there is the situation in which a group makes no judgment at all about whether the risks of a thing are acceptable or not—they simply do not think about it. By Lowrance's definition, this means the thing is neither safe nor unsafe with respect to that group. Yet this goes against our ordinary ways of thinking about safety. For example, we normally say that some cars are safe and others unsafe, even though many people may never even think about the safety of the cars they drive.

There must be at least some objective point of reference outside ourselves that allows us to decide whether our judgments about safety are correct once we have settled on what constitutes to us an acceptable risk. An expanded definition could capture this element, without omitting the insight already noted that safety judgments are relative to people's value perspectives.[3] One option is simply to equate safety with the absence of risk. Because little in life, and nothing in engineering, is risk-free, we prefer to adopt a modified version of Lowrance's definition:

> A thing is safe if, were its risks fully known, those risks would be judged acceptable by a reasonable person in light of settled value principles.

[3] Council for Science and Society, *The Acceptability of Risks* (England: Barry Rose, Ringwood, Hants, 1977), p. 13.

Safety is frequently thought of in terms of degrees and comparisons. We speak of something as "fairly safe" or "relatively safe" (compared with similar things). Using our definition, this translates as the degree to which a person or group, judging on the basis of their settled values, would decide that the risks of something are more or less acceptable in comparison with the risks of some other thing. For example, when we say that airplane travel is safer than automobile travel, we mean that for each mile traveled, it leads to fewer deaths and injuries—the risky elements that our settled values lead us to avoid. Finally, we interpret "things" to include products as well as services, institutional processes, and disaster protection.

Risks

We say a thing is not safe if it exposes us to unacceptable risk, but what is meant by "risk"? *A risk is the potential that something unwanted and harmful may occur.* We take a risk when we undertake something or use a product or substance that is not safe. William D. Rowe refers to the "potential for the realization of unwanted consequences from impending events."[4] Thus, a future, possible, occurrence of harm is postulated.

Risk, like harm, is a broad concept covering many different types of unwanted occurrences. In regard to technology, it can equally well include dangers of bodily harm, of economic loss, or of environmental degradation. These in turn can be caused by delayed job completion, faulty products or systems, and economically or environmentally injurious solutions to technological problems.

Good engineering practice has always been concerned with safety. But as technology's influence on society has grown, so has public concern about technological risks increased. In addition to measurable and identifiable hazards arising from the use of consumer products and from production processes in factories, some of the less obvious effects of technology are now also making their way to public consciousness. And while the latter are often referred to as "new risks," many of them have existed for some time. They are new only in the sense that (1) they are now identifiable (because of changes in magnitude of the risks they present, because they have passed a certain threshold of accumulation in our environment, or because our methods of identification and measurement have improved) or (2) the public's perception of them has changed (because of education, experience, media attention, or a reduction in other hitherto dominant and masking risks).

---

[4] William D. Rowe, *An Anatomy of Risk* (New York: Wiley, 1977), p. 24.

Meanwhile, natural hazards continue to threaten human populations. Technology has greatly reduced the scope of some of these, such as floods, but at the same time it has increased our vulnerability to other natural hazards, such as earthquakes, as they affect our ever-greater concentrations of population and cause greater damage to our finely tuned technological networks of long lifelines for water, energy, and food.

Acceptability of Risk

Having adopted a modified version of Lowrance's definition of safety as acceptable risk, we need to examine the idea of acceptability more closely. William D. Rowe says that "a risk is acceptable when those affected are generally no longer (or not) apprehensive about it."[5] Apprehensiveness depends to a large extent on how the risk is perceived. This is influenced by such factors as whether the risk is assumed voluntarily, the perception of the probabilities of harm (or benefit), the job-related or other pressures that cause people to be aware of or to overlook risks, the immediacy or obviousness of a risky activity or situation, and the identification of potential victims. Let us illustrate these elements of risk perception by means of some examples.

Voluntarism and Control

John and Ann Smith and their children enjoy riding motorcycles over rough terrain for amusement. They take voluntary risks, part of the thrill of being engaged in such a potentially dangerous sport. They do not expect the manufacturer of their dirt bikes to adhere to the same standards of safety as would the makers of a passenger car used for daily commuting. The bikes should be sturdy, but guards covering exposed parts of the engine, padded instrument panels, collapsible steering mechanisms, or emergency brakes are clearly unnecessary, if not inappropriate.

In discussing dirt bikes and the like, we do not include all-terrain three-wheel vehicles. Those represent hazards of greater magnitude because of the false sense of security they give the rider. They tip over easily. During the five years before they were forbidden in the United States, they were responsible for nearly 900 deaths and 300,000 injuries. About half of the casualties were children under 16.

John and Ann live near a chemical plant. It is the only area in which they can afford to live, and it is near the shipyard where they both work. At home, they suffer from some air pollution,

---

[5] William D. Rowe, "What Is an Acceptable Risk and How Can It Be Determined?" in *Energy Risk Management,* ed. G. T. Goodman and W. D. Rowe (New York: Academic, 1979), p. 328.

and there are some toxic wastes in the ground. Official inspectors tell them not to worry. Nevertheless, they do, and they think they have reason to complain—they do not care to be exposed to risks from a chemical plant with which they have no relationship except on an involuntary basis. Any beneficial link to the plant through consumer products or other possible connections is very remote and, moreover, subject to choice.

John and Ann behave as most of us would under the circumstances: We are much less apprehensive about the risks to which we expose ourselves voluntarily than those to which we are exposed involuntarily. In terms of our "engineering as social experimentation" paradigm, people are more willing to be the subjects of their own experiments (social or not) than of someone else's.

Intimately connected with this notion of voluntarism is the matter of control. The Smiths choose where and when they will ride their bikes. They have selected their machines and they are proud of how well they can control them (or think they can). They are aware of accident figures, but they tell themselves those apply to other riders, not to them. In this manner, they may well display the characteristically unrealistic confidence of most people when they believe hazards to be under their control.[6] But still, riding motor bikes, skiing, hang gliding, bungee jumping, horseback riding, boxing, and other hazardous sports are usually carried out under the assumed control of the participants. Enthusiasts worry less about their risks than the dangers of, say, air pollution or airline safety. Another reason for not worrying so much about the consequences of these sports is that rarely does any one accident injure any appreciable number of innocent bystanders.

Effect of Information on Risk Assessments

The manner in which information necessary for decision making is presented can greatly influence how risks are perceived. The Smiths are careless about using seat belts in their car. They know that the probability of their having an accident on any one trip is infinitesimally small. Had they been told, however, that in the course of 50 years of driving, at 800 trips per year, there is a probability of one in three that they will receive at least one disabling injury, then their seat belt habits (and their attitude

---

[6] Paul Slovic, Baruch Fischhoff, and Sarah Lichtenstein, "Weighing the Risks: Which Risks Are Acceptable?" *Environment* 21 (April 1979), pp. 14–20 and (May 1979), pp. 17–20, 32–38; Paul Slovic, Baruch Fischhoff, and Sarah Lichtenstein, "Risky Assumptions," *Psychology Today* 14 (June 1980), pp. 44–48.

about seat belt laws) would likely be different.[7] Studies have verified that a change in the manner in which information about a danger is presented can lead to a striking reversal of preferences about how to deal with that danger. Consider, for example, an experiment in which two groups of 150 people were told about the strategies available for combating a disease. The first group was given the following description:

> Imagine that the U.S. is preparing for the outbreak of an unusual Asian disease, which is expected to kill 600 people. Two alternative programs to combat the disease have been proposed. Assume that the exact scientific estimate of the consequences of the programs are as follows:
> If Program A is adopted, 200 people will be saved.
> If Program B is adopted, there is $1/3$ probability that 600 people will be saved, and $2/3$ probability that no people will be saved.
> Which of the two programs would you favor?[8]

The researchers reported that 72 percent of the respondents selected Program A, and only 28 percent selected Program B. Evidently the vivid prospect of saving 200 people led many of them to feel averse to taking a risk on possibly saving all 600 lives.

The second group was given the same problem and the same two options, but the options were worded differently:

> If Program C is adopted 400 people will die.
> If Program D is adopted, there is $1/3$ probability that nobody will die, and $2/3$ probability that 600 people will die.
> Which of the two programs would you favor?

This time only 22 percent chose Program C, which is the same as Program A, and 78 percent chose Program D, which is identical to Program B.

One conclusion that we draw from the experiment is that options perceived as yielding firm gains will tend to be preferred over those from which gains are perceived as risky or only probable. A second conclusion is that options emphasizing firm losses will tend to be avoided in favor of those whose chances of success are perceived as probable. In short, people tend to be more willing to take risks in order to avoid perceived firm losses than they are to win only possible gains.

---

[7] Richard J. Arnould and Henry Grabowski, "Auto Safety Regulation: An Analysis of Market Failure," *The Bell Journal of Economics* 12 (Spring 1981), p. 35.

[8] Amos Tversky and Daniel Kahneman, "The Framing of Decisions and the Psychology of Choice," *Science* 211 (January 30, 1981), p. 453.

Job-Related Risks

John Smith's work in the shipyard has in the past exposed him to asbestos. He is aware now of the high percentage of asbestosis cases among his co-workers, and, after consulting his own physician, finds that he is slightly affected himself. Even Ann, who works in a clerical position at the shipyard, has shown symptoms of asbestosis as a result of handling her husband's clothes. Earlier John saw no point to "all the fuss stirred up by some do-gooders." He figured that he was being paid to do a job; he felt the masks that were occasionally handed out gave him sufficient protection, and he thought the company physician was giving him a clean bill of health.

In this regard, John's thinking is similar to that of many workers who take risks on their jobs in stride, and sometimes even approach them with bravado. Of course, exposure to risks on a job is in a sense voluntary since one can always refuse to submit oneself to them, and workers perhaps even have some control over how their work is carried out. But often employees have little choice other than to stick with what is for them the only available job, and to do as they are told. What they are often not told about is their exposure to toxic substances and other dangers that cannot readily be seen, smelled, heard, or otherwise sensed. Unions and occupational health and safety regulations (such as right-to-know rules regarding toxics) can correct the worst situations, but standards regulating conditions in the workplace (its air quality, for instance) are generally still far below those that regulate conditions in our general (public) environment. It may be argued that the "public" encompasses many people of only marginal health whose low thresholds for pollution demand a fairly clean environment. On the other hand, factory workers are seldom carefully screened for their work. And in all but the most severe environments (those conducive to black lung or brown lung, for instance), unions have in the past displayed little desire for change, lest necessary modifications of the workplace force employers out of business altogether.

Engineers who design and equip work stations must take into account the cavalier attitude toward safety shown by many workers, especially when their pay is on a piecework basis. And when one worker complains about unsafe conditions but others do not, the complaint should not be dismissed as coming from a crackpot. Or consider the pain suffered by meat-cutters, or clerks at keyboards, whose afflictions by the carpal tunnel syndrome are still not universally recognized as debilitating in terms of workers' compensation (only recently did the Occupational Health and Safety Administration get political support to draft further ergonomic rules). All reports from the workplace regarding unsafe conditions merit serious attention by engineers.

Magnitude and Proximity

Our reaction to risk is affected by the dread of a possible mishap, both in terms of its magnitude and of how we may be related to, or identify with, the potential victims. A single major airplane crash in a remote country, the specter of a child we know or observe on the television screen trapped in a cave-in—these affect us more acutely than the ongoing but anonymous carnage on the highways, at least until someone close to us is involved in a car accident.

In terms of numbers alone, we feel much more keenly about a potential risk if one of us out of a group of 20 intimate friends is likely to be subjected to great harm than if it might affect, say, 50 strangers out of a proportionally larger group of 1000. This proximity effect arises in perceptions of risk over time as well. A future risk is easily dismissed by various rationalizations including (1) the attitude of "out of sight, out of mind," (2) the assumption that predictions for the future must be discounted by using lower probabilities of occurrences, or (3) the belief that a countermeasure will be found in time.

Misperceptions of numbers can easily make us overlook losses that are far greater than the numbers reveal by themselves. Consider the 75 men lost when the unfinished Quebec Bridge collapsed in 1907. As William Starna relates,

> Of those 75 men, no fewer than 35 were Mohawk Indians from the Caughnawaga Reserve in Quebec. Their deaths had a devastating effect on this small Indian community, altering drastically its demographic profile, its economic base, and its social fabric. Mohawk steelworkers would never again work in such large crews, opting instead to work in small groups on several jobs.[9]

Engineers face two problems with public conceptions of safety. On the one hand, there is the overly optimistic attitude that things that are familiar, that have not hurt us before, and over which we have some control present no real risks. On the other hand is the dread people feel when an accident kills or maims in large numbers, or harms those we know, even though statistically speaking such accidents might occur infrequently.

Leaders of industry are sometimes heard to proclaim that those who fear the effects of air pollution, toxic wastes, or nuclear power are emotional and irrational, or politically motivated. This in our view is a misperception of legitimate concerns expressed

---

[9] William A. Starna, "A Disaster's Toll," letter to the editor, *American Heritage of Invention and Technology,* Summer 1986, commenting on "A Disaster in the Making" in the Spring 1986 issue.

publicly by thoughtful citizens. It is important that engineers recognize as part of their work such widely held perceptions of risk and take them into account in their designs.

Discussion Topics

1. Describe a real or imagined traffic problem in your neighborhood involving children and elderly people who find it difficult to cross a busy street. Put yourself in the position of (*a*) a commuter traveling to work on that street; (*b*) the parent of a child or the relative of an older person who has to cross that street on occasion; (*c*) a police officer assigned to keep the traffic moving on that street; and (*d*) the town's traffic engineer working under a tight budget.

    Describe how in these various roles you might react to (*e*) complaints about conditions dangerous to pedestrians at that crossing and (*f*) requests for a pedestrian crossing protected by traffic or warning lights.

2. In some technologically advanced nations, a number of industries that have found themselves restricted by safety regulations have resorted to dumping their products on—or moving their production processes to—less-developed countries where higher risks are tolerated. Examples are the dumping of unsafe or ineffective drugs on the Third World by pharmaceutical companies from highly industrialized countries, and in the past the transfer of asbestos processing from the United States to Mexico.[10] More recently, toxic wastes (from lead-acid batteries to nuclear wastes) have been added to the list of "exports." To what extent do differences in perception of risk justify the transfer of such merchandise and production processes to other countries? Is this an activity that can or should be regulated?

3. Grain dust is pound for pound more explosive than coal dust or gunpowder. Ignited by an electrostatic discharge or other cause, it has ripped apart grain silos and killed or wounded many workers over the years. When 54 people were killed during Christmas week 1977, grain handlers and the U.S. government finally decided to combat dust accumulation.[11] Ten years, 59 deaths, and 317 serious injuries later, a compromise standard was agreed on that designates dust accumulation of $1/8$ inch or more

---

[10] Milton Silverman, Philip Lee, and Mia Lydecker, *Prescription for Death: The Drugging of the Third World* (San Francisco: University of California Press, 1981); and Henry Shue, "Exporting Hazards," *Ethics* 91 (July 1981), p. 586.

[11] Eliot Marshall, "Deadlock over Explosive Dust," *Science* 222 (November 4, 1983), pp. 485–87; discussion, p. 1183.

as dangerous and impermissible. Use grain facility explosions for a case study of workplace safety and rule making.

Assessing and Reducing Risk

Any improvement in safety as it relates to an engineered product is often accompanied by an increase in the cost of that product. On the other hand, products that are not safe burden the manufacturer with secondary costs beyond the primary (production) costs—costs associated with warranty expenses, loss of customer goodwill and even loss of customers because of injuries sustained from use of the product, litigation, possible downtime in the manufacturing process, and so forth. (See Figure 4–1.) It is therefore important for manufacturers and users alike to reach some understanding of the risks connected with any given product and know what it might cost to reduce those risks (or not reduce them).

Uncertainties in Design

One would think that experience and historical data would provide good information about the safety of standard products.

Figure 4–1
Why both low-risk and high-risk products are costly. P = primary cost of product, including cost of safety measures involved; S = secondary costs, including warranties, loss of customer goodwill, litigation costs, costs of down time, and other secondary costs. T = total cost. Minimum total cost occurs at M, where incremental savings in primary cost (slope of P) are offset by an equal incremental increase in secondary cost (slope of S). Highest acceptable risk (H) may fall below risk at least cost (M), in which case H and its higher cost must be selected as the design or operating point.

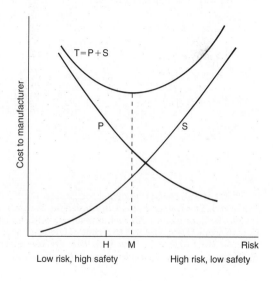

Much has been collected and published; gaps remain, however, because (1) there are some industries where information is not freely shared, for instance when the cost of failure is less than the cost of fixing the problem, (2) problems and their causes are not revealed after a legal settlement has been reached, and (3) there are always new applications of old technology, or substitution of materials and components, that render the available information less useful.

Risk is seldom intentionally designed into a product. It arises because of the many uncertainties faced by the design engineer, the manufacturing engineer, and even the sales and applications engineer.

To start with, there is the purpose of a design. Let us consider an airliner. Is it meant to maximize profits for the airline, or is it intended to give the highest possible return on investment? The answer to that question is important to the company because on it hinge a number of decisions and their outcomes and the possibility of the airline's economic success or ruin. Investing $50 million in a jet to bring in maximum profits of, say, $10 million during a given time involves a lower return on investment than spending $24 million on a medium-sized jet to bring in a return of $6 million in that same period.

Regarding applications, designs that do quite well under static loads may fail under dynamic loading. A historical example is the wooden bridge that shook and collapsed when a contingent of Napoleon's army crossed it marching in step. Such vibrations even affected one of Robert Stephenson's steel bridges, which swayed violently under a contingent of marching British troops. Ever since then, soldiers are under orders to fall out of step when crossing a bridge. Wind can also cause destructive vibrations. Two examples are (1) "Galloping Gertie"—the widely discussed but still not completely explained collapse of the Tacoma Narrows Bridge[12] in 1940—and (2) a high voltage power line across the Bosporus in Turkey. When aerial cables of this power line oscillated during a strong wind, arcing melted them where they touched and they fell on houses and people below.

Apart from uncertainties about applications of a product, there are uncertainties regarding the materials of which it is made and the level of skill that goes into designing and manu-

---

[12] Regarding the Tacoma Narrows Bridge, see Henry Petroski, *To Engineer Is Human: The Role of Failure in Successful Design* (New York: St. Martin's Press, 1985), or M. Levy and M. Salvadori, *Why Buildings Fall* (New York: C. C. Norton & Co., 1992). For recent challenges to earlier explanations of the collapse, see Frederic D. Schwarz, "Why Theories Fall Down," *American Heritage of Invention & Technology,* Winter 1993, pp. 6–7.

facturing it. For example, changing economic realities or hitherto unfamiliar environmental conditions such as extremely low temperatures may affect how a product is to be designed. A typical "handbook engineer" who extrapolates tabulated values without regard to their implied limits under different conditions will not fare well under such circumstances.

Caution is required even with standard materials specified for normal use. In 1981, a new bridge that had just replaced an old and trusted ferry service across the Mississippi at Praire du Chien, Wisconsin, had to be closed because 11 of the 16 flange sections in both tie girders were found to have been fabricated from excessively brittle steel.[13] (In the meantime, the ferries had disappeared!) While strength tests are (supposedly) routinely carried out on concrete, the strength of steel is all too often taken for granted.

Such drastic variations from the standard quality of a given grade of steel are exceptional; more typically the variations are small. Nevertheless, the design engineer should realize that the supplier's data on items such as steel, resistors, insulation, optical glass, and so forth apply to statistical averages only. Individual components can vary considerably from the mean. Engineers traditionally have coped with such uncertainties about materials or components, as well as incomplete knowledge about the actual operating conditions of their product, by introducing a comfortable "factor of safety." That factor is intended to protect against problems arising when the stresses due to anticipated loads (duty distribution) and the stresses the product as designed is supposed to withstand (capability distribution) depart from their respective expected values. Stresses can be of a mechanical or other nature—for example, an electric field gradient to which an insulator is exposed or the traffic density at an intersection.

A product may be said to be safe if its capability exceeds its duty. But this presupposes exact knowledge of actual capability and actual duty. In reality the stress calculated by the engineer for a given condition of loading and the stress that ultimately materializes at that loading may vary quite a bit. This is because each component in an assembly has been allowed certain tolerances in its physical dimensions and properties—otherwise the production cost would be prohibitive. The result is that the assembly's capability as a whole cannot be given by a single numerical value but must be expressed as a probability density that can be graphically depicted as a "capability" curve (Figure

---

[13] "Suit Claims Faulty Bridge Steel," *ENR* (*Engineering News Record*), March 12, 1981, p. 14; see also March 26, 1981, p. 20; April 23, 1981, pp. 15–16; November 19, 1981, p. 28.

Figure 4–2
Probability density curves
for stress in an
engineered system.
(a) Variability of stresses
in a relatively safe case.
(b) Lower safety due to
overlap in stress
distributions.

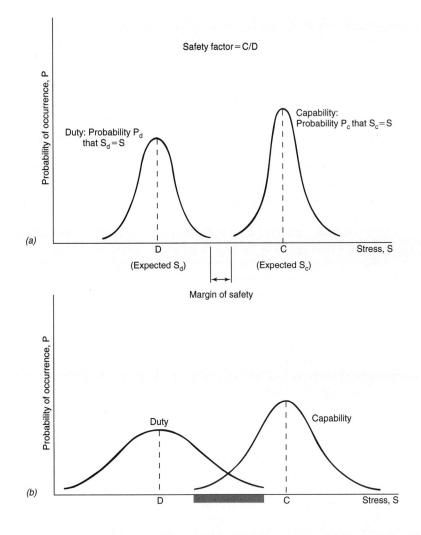

4–2). For a given point on a capability curve, the value along the
vertical axis gives the probability that the capability, or strength,
is equal to the corresponding value along the horizontal axis.

A similar curve can be constructed for the duty that the
assembly will actually experience. The stress exposure varies
because of differences in loads, environmental conditions, or the
manner in which the product is used. Associated with the capability and duty curves are nominal or, statistically speaking,
expected values C and D. We often think and act only in terms of
nominal or expected values. And with such a deterministic frame
of mind, we may find it difficult to conceive of engineering as
involving experimentation. The "safety factor" C/D rests com-
fortably in our consciences. But how sure can we be that our

materials are truly close to their specified nominal properties? Or that the loads will not vary too widely from their anticipated values or occur in environments hostile to the proper functioning of the materials? It is entirely conceivable that the capability and duty curves will assume flatter shapes because of increased variances (see Figure 4–2b) than they would have under normal conditions, as in Figure 4–2a. And Figure 4–2b shows how it is probable that load stress may exceed design stress along the shaded region of stress. Mathematical treatment of this topic is offered by, among others, Edward B. Haugen, who reminds his readers how "the safety factor concept completely ignores the facts of variability that result in different reliabilities for the same safety factor."[14]

A more appropriate measure of safety would be the "margin of safety," which is shown in Figure 4–2a. If it is difficult to compute such a margin of safety for ordinary loads used every day, imagine the added difficulties that arise when repeatedly changing loads have to be considered.

Risk–Benefit Analyses

Many large projects, especially public works, are justified on the basis of a risk–benefit analysis. The questions answered by such a study are the following: Is the product worth the risks connected with its use? What are the benefits? Do they outweigh the risks? We are willing to take on certain levels of risk as long as the project (the product, the system, or the activity that is risky) promises sufficient benefit or gain. If risk and benefit can both be readily expressed in a common set of units (say, lives or dollars as in cost-benefit analyses), it is relatively easy to carry out a risk–benefit analysis and to determine whether we can expect to come out on the benefit side. For example, an inoculation program may produce some deaths, but it is worth the risk if many more lives are saved by suppressing an imminent epidemic.

A closer examination of risk–benefit analysis reveals some conceptual difficulties. Both risks and benefits lie in the future. Since there is some uncertainty associated with them, we should address their expected values (provided such a model fits the situation); in other words, we should multiply the magnitude of the potential loss by the probability of its occurrence, and similarly with the gain. But who establishes these values, and how? If the benefits are about to be realized in the near future but the risks are far off (or vice versa), how is the future to be discounted in terms of, say, an interest rate so we can compare present values?

---

[14] Edward B. Haugen, *Probabilistic Approaches to Design* (New York: Wiley, 1968), p. 5.

What if the benefits accrue to one party and the risks are incurred by another party?

The matter of delayed effects presents particular difficulties when an analysis is carried out during a period of high interest rates. Under such circumstances, the future is discounted too heavily because the very low present values of cost or benefit do not give a true picture of what a future generation will face.

How should one proceed when risks or benefits are composites of ingredients that cannot be added in a common set of units, as for instance in assessing effects on health plus aesthetics plus reliability? At most, one can compare designs that satisfy some constraints in the form of "dollars not to exceed X, health not to drop below Y" and attempt to compare aesthetic values with those constraints. Or when the risks can be expressed and measured in one set of units (say, deaths on the highway) and benefits in another (speed of travel), we can employ the ratio of risks to benefits for different designs when comparing the designs.

It should be noted that risk–benefit analysis, like cost–benefit analysis, is concerned with the advisability of undertaking a *project*. When we judge the relative merits of different *designs,* however, we move away from this concern. Instead we are dealing with something similar to cost-effectiveness analysis, which asks what design has the greater merit—given that the project is actually to be carried out. Sometimes the shift from one type of consideration to the other is so subtle that it passes unnoticed. Nevertheless, engineers should be aware of the differences so that they do not unknowingly carry the assumptions behind one kind of concern into their deliberations over the other.

These difficulties notwithstanding, there is a need in today's technological society for some commonly agreed on process—or at least a process open to scrutiny and open to modification as needed—for judging the acceptability of potentially risky projects. What we must keep in mind is the following ethical question: "Under what conditions, if any, is someone in society entitled to impose a risk on someone else on behalf of a supposed benefit to yet others?"[15] Here we must not restrict our thoughts to average risks and benefits, but should also consider those worst-case scenarios of persons exposed to maximum risks while they are also reaping only minimum benefits. Are their rights violated? Are they provided safer alternatives? In examining this problem further, we should also trace our steps back to an observation on risk perception made earlier: A risk to a known person (or to identifiable individuals) is perceived differently from sta-

---

[15] Council for Science and Society, *The Acceptability of Risks,* p. 37.

tistical risks merely read or heard about. What this amounts to
is that engineers do not affect just an amorphous public; their
decisions have a direct impact on people who feel the impact
acutely, and that fact should be taken into account equally as
seriously as are studies of statistical risk.

Personal Risk

Given sufficient information, an individual is able to decide
whether to participate in (or consent to exposure to) a risky
activity (an experiment). Chauncey Starr has prepared some
widely used figures that indicate that individuals are more ready
to assume voluntary risks than they are when subjected to involuntary risks (or activities over which they have no control), even
when the voluntary risks are 1000 times more likely to produce
a fatality than the involuntary ones (Figure 4–3).

The difficulty in assessing personal risks arises when we consider those that are involuntary. Take John and Ann Smith and
their discomfort over living near a refinery. Assume the general
public was all in favor of building a new refinery at that location,
and assume the Smiths already lived in the area. Would they
and others in their situation have been justified in trying to veto
its construction? Would they have been entitled to compensation
if the plant was built over their objections anyway? If so, how

Figure 4–3
Willingness to assume voluntary risks as opposed to involuntary ones
correlated to benefits those risks produce. (*Adapted from C. Starr, "Social
Benefit versus Technological Risk,"* Science *165: 1232–38.*)

Reprinted with permission from "Social Benefit versus Technological Risk," Science *165:*
1232–38. Copyright 1969 American Association for the Advancement of Science.

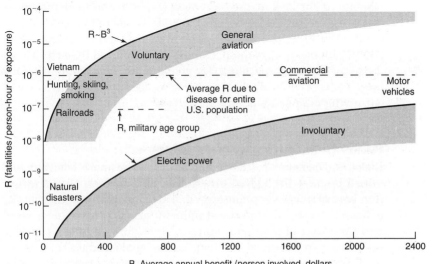

much compensation would have been adequate? These questions arise in numerous instances. Nuclear power plant siting is another example. Indeed, Figure 4–3 was produced in the context of nuclear safety studies.

The problem of quantification alone raises innumerable problems in assessing personal safety and risk. How, for instance, is one to assess the dollar value of an individual's life? This question is as difficult as deciding whose life is worth saving, should such a choice ever have to be made.

Some would advocate that the marketplace should decide, assuming market values can come into play. But today there is no over-the-counter trade in lives. Nor are even more mundane gains and losses easily priced. If the market is being manipulated, or if there is a wide difference between "product" cost and sales price, it matters under what conditions the buying or selling takes place. For example, if one buys a loaf of bread, it can matter whether it is just one additional daily loaf among others one buys regularly; it is different when it is the first loaf available in weeks. Or, if you are compensated for a risk by an amount based on the exposure tolerance of the average person, yet your tolerance of a condition or your propensity to be harmed is much greater than average, the compensation is apt to be inadequate.

The result of these difficulties in assessing personal risk is that analysts employ whatever quantitative measures are ready at hand. In regard to voluntary activities, one could possibly make judgments on the basis of the amount of life insurance taken out by an individual. Is that individual going to offer the same amount to a kidnapper to be freed? Or is there likely to be a difference between future events (requiring insurance) and present events (demand for ransom)? In assessing a hazardous job, one might look at the increased wages a worker demands to carry out the task. Faced with the wide range of variables possible in such assessments, one can only suggest that an open procedure, overseen by trained arbiters, be employed in each case as it arises. On the other hand, for people taken in a population-at-large context, it is much easier to use statistical averages without giving offense to anyone in particular.

Public Risk and Public Acceptance

Risks and benefits to the public at large are more easily determined because individual differences tend to even out as larger numbers of people are considered. The contrast between costs of a disability viewed from the standpoint of a private value system and from that of a societal value system, for example, is vividly illustrated in Figure 4–4. Also, assessment studies relating to technological safety can be conducted more readily in the detached manner of a macroscopic view as statistical parameters

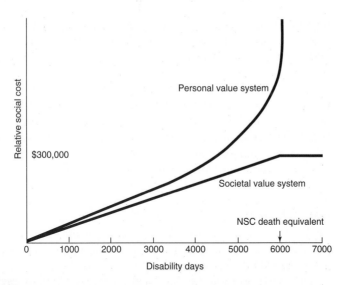

Figure 4–4
Value systems for social costs of disability. Using the National Safety
Council equivalent of 6000 disability days for death and L. A. Sagan's
1972 assumed rate of $50 per day of disability (Sagan, "Human Cost of
Nuclear Power," *Science* 177, pp. 487–93) yields a "death equivalent" of
$300,000—valid for societal value analysis only. (*Adapted from Starr,
Rundman, and Whipple, "Philosophical Basis for Risk Analysis,"* Annual
Review of Energy 1, pp. 629–62).

With permission from the *Annual Review of Energy and Environment,* vol. 1, © 1976 by Annual
Reviews, www.annualreviews.org

take on greater significance. In that context, the National Highway Traffic Safety Administration (NHTSA) has proposed a value
for human life based on loss of future income and other costs
associated with an accident. Intended for study purposes only,
NHTSA's "blue-book value" amounted to $200,725 in 1972 dol-
lars. This is certainly a more convenient measure than sorting
out the latest figures from court cases. (On April 23, 1981, for
example, the *Los Angeles Times* reported settlements for rela-
tives of victims of the 1979 DC-10 crash in Chicago at $2,287,257
for a 36-year-old promising executive, $750,000 for a telephone
company employee, and $275,000 for a stewardess on duty.) In a
more recent case the family of a mechanic killed in an automobile
accident was awarded $6 million for the loss of the breadwinner.
Punitive damages amounted to an additional $20 million.

Shulamit Kahn finds a labor market value of life in the
amount of $8 million to be an acceptable average value to many
individuals questioned by various investigators. Interestingly,
people place higher values on *other* persons' lives, on the order of
115 to 230 percent more. Kahn says, "Yet even the $8 million fig-
ure is higher than is typically used in policy analysis. The
unavoidable implication . . . is therefore that policy analysts do

not evaluate the risk of their subjects' lives as highly as people evaluate risks to their own (and others') lives. Consequently, too many risks are taken."[16]

NHTSA, incidentally, emphasized that "placing a value on a human life can be nothing more than a play with figures. We have provided an estimate of some of the quantifiable losses in social welfare resulting from a fatality and can only hope that this estimate is not construed as some type of basis for determining the 'optimal' (or even worse, the 'maximum') amount of expenditure to be allocated to saving lives."[17]

Some Examples of Improved Safety

This is not a treatise on design; therefore, only a few simple examples will be given to show that safety need not rest on elaborate contingency features.

The first example is the magnetic door catch introduced on refrigerators to prevent death by asphyxiation of children accidentally trapped in them. The catch in use today permits the door to be opened from the inside without major effort. It also happens to be cheaper than the older types of latches.

The second example is the dead-man handle used by the engineer (engine driver) to control a train's speed. The train is powered only as long as some pressure is exerted on the handle. If the engineer becomes incapacitated and lets go of the handle, the train stops automatically. Perhaps cruise controls for newer-model automobiles should come equipped with a similar feature.

Railroads provide the third example as well. Over a hundred years ago, to signal to a train that it could proceed, a ball was raised to the top of a mast (hence the term "highball" for full speed ahead); to signal a stop, the ball was lowered. Later, semaphores used a mechanical arm, but both methods required a cable to be pulled and incorporated a fail-safe approach in that an accidentally cut cable would let the ball or the arm drop to the STOP position all by itself.

The motor-reversing system shown in Figure 4–5 gives still another example of a situation in which the introduction of a safety feature involves merely the proper arrangement of functions at no additional expense. As the mechanism is designed in Figure 4–5a, sticky contacts could cause battery B to be shorted

---

[16] Shulamit Kahn, "Economic Estimates of the Value of Life," *IEEE Technology and Society Magazine,* June 1986, pp. 24–29. Reprinted in Albert Flores, *Ethics and Risk Management in Engineering* (Boulder, CO: Westview Press, 1988).

[17] Brian O'Neill and A. B. Kelley, "Costs, Benefits, Effectiveness, and Safety: Setting the Record Straight," *Professional Safety,* August 1975, p. 30.

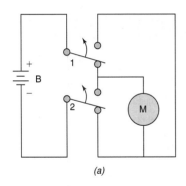 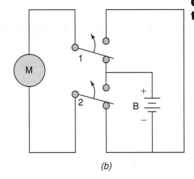

(a)   (b)

Figure 4–5
Reversing switch for a permanent magnet motor. (a) Arms 1 and 2 of the
switch are both raised by a solenoid (not shown). If either one does not
move—say, a contact sticks—while the other does, there is a short across
the battery. The battery will discharge and be useless even after the
trouble is detected. (b) By exchanging the positions of battery and motor,
a stuck switch will cause no harm to the battery. (The motor can be
shorted without harm.)

and discharged, thus making it unavailable for further use even
after the contacts are coaxed loose. A simple reconnection of
wires as shown in Figure 4–5b removes that problem altogether.

In the rush to bring a product onto the market, safety considerations are frequently slighted. This would not be so much the
case if the venture were regarded as an experiment—an experiment that is about to enter its active phase as the product comes
into the hands of the user. Space flights were carried out with
such an attitude, but more mundane ventures involve less obvious dangers and therefore less attention is usually paid to safety.
If moral concerns alone do not sway engineers and their employers to be more heedful of potential risks, then the threat of litigation should certainly do so. However, this is not altogether
helpful to the public because many cases brought to court are
eventually settled and the findings sealed, depriving the public
and other manufacturers of information relevant to making
informed choices.

Discussion Topics

1. A worker accepts a dangerous job after being offered an annual
    bonus of $2000. The probability that the worker may be killed in
    any one year is 1 in 10,000. This is known to the worker. The
    bonus may therefore be interpreted as a self-assessment of life
    with a value equal to $2000 divided by 1/10,000, or $20 million.
    Is the worker more or less likely to accept the job if presented
    with the statistically nearly identical figures of a $100,000 bonus

over 50 years (neglecting interest) and a $1/200$ probability of a fatal accident during that period?

2. "Airless" paint spray guns do not need an external source of compressed air connected to the gun by a heavy hose (although they do need a cord to attach them to a power source) because they have incorporated a small electric motor and pump. One common design uses an induction motor that does not cause sparking since it does not require a commutator and brushes (which are sources of sparking). Nevertheless, the gun carries a label, warning users that electrical devices operated in paint spray environments pose special dangers. Another type of gun that, like the first, also requires only a power cord is designed to weigh less by using a high-speed universal motor and a disk-type pump. The universal motor does require a commutator and brushes, which cause sparking. This second kind of spray gun carries a warning similar to that attached to the first, but it states in addition that the gun should never be used with paints that employ highly volatile and flammable thinners such as naphtha. The instruction booklet is quite detailed in its warnings.

   A painter had been lent one of the latter types of spray guns. In order to clean the apparatus, he partially filled it with paint thinner and operated it. It caught fire, and the painter was severely burned as the fire spread. The instruction booklet was in the cardboard box in which the gun was kept, but it had not been read by the painter, who was a recent immigrant and did not read English very well. He, however, had used the first type of airless paint spray gun in a similar manner without mishap. The warning messages on both guns looked pretty much the same. Do you see any ethical problems in continuing over-the-counter sales of this second type of spray gun? What should the manufacturer of this novel, lightweight device do?

   In answering these questions, consider the fact that courts have ruled that hidden design defects are not excused by warnings attached to the defective products or posted in salesrooms. Informed consent must rest on a more thorough understanding than can be transmitted to buyers by warning labels.

3. During the early stages of its development, crash tests revealed that Ford's *Pinto* could not sustain a front-end collision without the windshield breaking. So the drive train was moved backward until the gas tank was close to the differential. Once the *Pinto* became available to the public in the early '70s, a number of rear-end collisions led to serious gas-tank explosions and subsequent law-suits. Study the Pinto story and ask yourself when, and in what sense, a price can be put on a human life.[18]—Fur-

---

[18] Frank Camps, "Warning an Auto Company about an Unsafe Design," in *Whistle-Blowing!: Loyalty and Dissent in the Corporation,* ed. Alan F. Westin (New York: McGraw-Hill, 1981), pp. 119–29. See also Douglas Birsch and John

thermore, why didn't GM learn from Ford's experience? The gas tank of a '79 Chevrolet *Malibu,* too close to the bumper, exploded in a 1993 rear-end collision with fatal results. A GM engineer testified that moving the tank forward and shielding it would have cost $8.59 per car against $2.40 per car in possible damages averaged over all cars of the product line. In August 1999 a jury awarded the Patricia Anderson family $107 Million in compensatory damages and $4.8 Billion in punitive damages (subject to appeal) for the 1993 crash.

Three Mile Island, Chernobyl, and Citicorp Tower

As our engineered systems grow more complex, it becomes more difficult to operate them. As Charles Perrow[19] argues, our traditional systems tended to incorporate sufficient slack, which allowed system aberrations to be corrected in a timely manner. Nowadays, he points out, subsystems are so tightly coupled within more complex total systems that it is not possible to alter a course safely unless it can be done quickly and correctly. Frequently, the supposedly corrective action taken by operators may make matters worse because they do not know what the problem is. For instance, during the emergency at Three Mile Island (to be described next), so many alarms had to be recorded by a printer that it fell behind by as much as 2 $\frac{1}{2}$ hours in displaying the events.

Designers hope to ensure greater safety during emergencies by taking human operators out of the loop and mechanizing their functions. The control policy would be based on predetermined rules. This in itself creates problems because (1) not all eventualities are foreseeable and (2) even those that can be predicted will be programmed by an error-prone human designer. In addition, another problem arises when the mechanized system fails and a human operator has to replace the computer in an operation that demands many rapid decisions.

Operator errors were the main cause of the nuclear reactor accidents at Three Mile Island and Chernobyl. Beyond these errors, a major deficiency surfaced at both installations: inadequate provisions for evacuation of nearby populations. This lack of "safe exit" is found in too many of our amazingly complex systems.

The third case we will examine shows how cooperation among designer, builder, and user of a major structure such as a skyscraper is important beyond construction as part of ongoing monitoring.

H. Fielder, eds. *The Ford Pinto Case* (Albany, NY: State University of New York Press, 1994). For wider discussions of risk–benefit analysis, see Kristin S. Shrader-Frechette, *Risk Analysis and Scientific Method* (Dordrecht, Netherlands: D. Reidel, 1985) and C. E. Harris et al, *Engineering Ethics,* p. 184.

[19] Charles Perrow, *Normal Accidents: Living with High-Risk Technologies* (New York: Basic Books, 1984).

Three Mile Island

Walter Creitz, president of Metropolitan Edison, the power company in the Susquehanna Basin, was obviously annoyed by a series of articles in the *Record,* a local, daily newspaper of York, Pennsylvania. The *Record* had cited unsafe conditions at Metropolitan Edison's Three Mile Island nuclear power plant Unit 2. Creitz dismissed the stories as "something less than a patriotic act—comparable in recklessness . . . to shouting 'Fire!' in a crowded theater." A few days later, a minor malfunction in the plant set off a series of events that made "Three Mile Island" into household words across the world.[20]

Briefly, this is what happened.[21] At 4 A.M. on March 28, 1979, Unit TMI-2 was operating under full automatic control at 97 percent of its rated power output. For 11 hours, a maintenance crew had been working on a recurring minor problem. Resin beads are used in several demineralizers (labeled 14 in Figure 4–6) to clean or "polish" the water on its way from the steam condenser (12) back to the steam generator (3). Some beads clogged the resin pipe from a demineralizer to a tank in which the resin is regenerated. In flushing the pipe with water, perhaps a cupful of water backed up into an air line that provides air for fluffing the resin in its regeneration tank. But that air line is connected to the air system that also serves the control mechanisms of the large valves at the outlet of the demineralizers. Thus, it happened that these valves closed unexpectedly without warning.

With water flow interrupted in the secondary loop (26), all but one of the condensate booster pumps turned off. That caused the main feedwater pumps (23) and the turbine (10) to shut down as well. In turn, an automatic emergency system started up the auxiliary feedwater pumps (25). But with the turbines inoperative, there was little outlet for the heat generated by the fission process in the reactor core. The pressure in the reactor rose to over 2200 pounds per square inch, opening a pressure-relief

---

[20] Mitchell Rogovin and George T. Frampton Jr., *Three Mile Island: A Report to the Commissioners and the Public,* vol. 1, Nuclear Regulatory Commission Special Inquiry Group, NUREG/CR-1250, Washington, DC (January 1980), p. 3. Diagram in text used with permission of Mitchell Rogovin.

[21] For details, see Kemeny Commission Report, *Report of the President's Commission on the Accident at Three Mile Island* (New York: Pergamon Press, 1979); Daniel F. Ford, *Three Mile Island* (New York: Viking, 1982); John F. Mason, "The Technical Blow-by-Blow: An Account of the Three Mile Island Accident," *IEEE Spectrum* 16 (November 1979), pp. 33–42; Thomas H. Moss and David L. Sills, eds., *The Three Mile Island Nuclear Accident: Lessons and Implications,* vol. 365 (New York: Annals of the New York Academy of Sciences, New York, 1981); Bill Keisling, *Three Mile Island* (Seattle, WA: Veritas, 1980); Daniel Martin, *Three Mile Island: Prologue or Epilogue?* (Cambridge, MA: Ballinger, 1980).

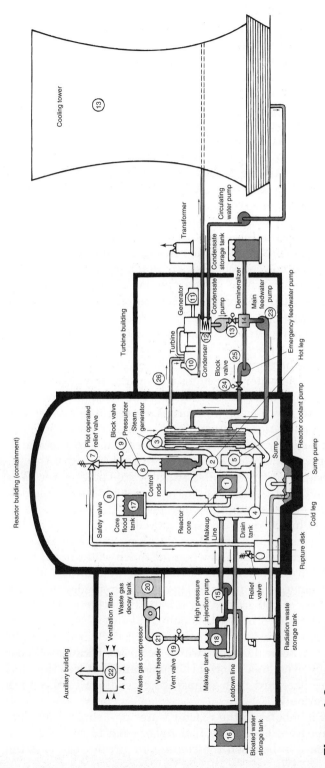

Figure 4–6

Schematic diagram of Three Mile Island nuclear power plant Unit 2. Pressurized water reactor system: Heat from reactor core (1) is carried away by water in a primary loop (1, 2, 3, 5, 4). In steam generator (3), the heat is transferred to water in a secondary loop (26) at lower pressure. The secondary-loop water turns to steam in the steam generator or boiler (3), drives the turbine (10), turns into water in the condenser (12), and is circulated back to (3) by means of pumps (13, and 23 and 25). (*Adapted from John F. Mason, "The Technical Blow-by-Blow: An Account of the Three Mile Island Accident," IEEE Spectrum, 16 [November 1979], copyright © 1979 by the Institute of Electrical and Electronics Engineers, Inc., and from Mitchell Rogovin and George T. Frampton Jr., Three Mile Island: A Report to the Commissioners and the Public, vol. 1, Nuclear Regulatory Commission Special Inquiry Group, NUREG/CR=1250, Washington, DC [January 1980].*)

valve (7) and signaling a SCRAM, in which control rods are lowered into the reactor core to stop the main fission process.

The open valve succeeded in lowering the pressure, and the valve was readied to be closed. Its solenoid was de-energized and the operators were so informed by their control-panel lights. But something went wrong: The valve remained open, contrary to what the control panel indicated. Apart from this failure, everything else had proceeded automatically as it was supposed to. Everything, that is, except for one other serious omission: The auxiliary pumps (25) that had been started automatically could not supply the auxiliary feedwater because block valves 24 had inadvertently been left closed after maintenance work done on them two days earlier. Without feedwater in the loop (26), the steam generator (3) boiled dry. Now there was practically no heat removal from the reactor, except through the relief valve. Water was pouring out through it at the rate of 220 gallons per minute. The reactor had not yet cooled down, and even with the control rods shutting off the main fission reaction, there would still be considerable heat produced by the continuing radioactive decay of waste products.

Loss of water in the reactor caused one of a group of pumps, positioned at 15 in the auxiliary building, to start automatically; another one of these pumps was started by the operators to rapidly replenish the water supply for the reactor core. Soon thereafter, the full emergency core-cooling system went into operation in response to low reactor pressure. Low reactor pressure can promote the formation of steam bubbles that reduce the effectiveness of heat transfer from the nuclear fuel to the water. There is a pressurizer that is designed to keep the reactor water under pressure. (The relief valve sits atop this pressurizer.) The fluid level in the pressurizer is also used as an indirect—and the only—means of measuring the water level in the reactor.

The steam in the reactor vessel caused the fluid level in the pressurizer to rise. The operators, thinking they had resolved the problem and that they now had too much water in the reactor, shut down the emergency core-cooling system and all but one of the emergency pumps. Then they proceeded to drain water at a rate of 160 gallons per minute from the reactor, causing the pressure to drop. At this point, they were still unaware of the water escaping through the open relief valve. Actually, they assumed some leakage, which occurred because of poor valve seating even under normal circumstances. It was this that made them disregard high-temperature readings in the pipes (beyond location 7).

The steam bubbles in the reactor water now covered much of the fuel, and the tops of the fuel rods began to crumble. The chemical reaction between the steam and zircaloy covering the fuel elements produced hydrogen, some of which was released into the containment structure, where it exploded.

The situation was becoming dire when, two hours after the initial event, the next shift arrived for duty. With some fresh insights into the situation, the relief valve was deduced to be open. Blocking valve (9) in the relief line was then closed by the new crew. Soon thereafter, with radiation levels in the containment building rising, a general alarm was sounded. While there had been telephone contact with the Nuclear Regulatory Commission (NRC), as well as with Babcock & Wilcox (B&W), who had constructed the reactor facility, no one answered at NRC's regional office and a message had to be left with an answering service. The fire chief of nearby Middletown was to hear about the emergency on the evening news.

In the meantime, a pump was transferring the drained water from the main containment building to the adjacent auxiliary building, but not into a holding tank as intended; because of a blown rupture disk, the water landed on the floor. When there was indication of sufficient airborne radiation in the control room to force evacuation, all but essential personnel wearing respirators stayed behind. The respirators made communication difficult.

Eventually the operators decided to turn the high-pressure injection pumps on again, as the automatic system had been set to do all along. The core was covered once more with water, though there were still some steam and hydrogen bubbles on the loose. Thirteen and one-half hours after the start of the episode there was finally hope of getting the reactor under control. Confusion over the actual state of affairs, however, continued for several days.

Nationwide the public watched television coverage in disbelief as responsible agencies displayed their lack of emergency preparedness at both the reactor site and evacuation-planning centers. Years later one still reads about the steadily accumulating costs of decommissioning (defueling, decontaminating, entombing) Unit 2 at TMI, $1 billion so far, of which 1/3 is passed on to ratepayers—all this for what cost $700 million to build. Three Mile Island was a financial disaster and a blow to the reputation of the industry, but fortunately radioactive release was low and cancer rates downwind are reported to be only slightly higher than normal.

Chernobyl

The nuclear power plant complex at Chernobyl, near Kiev (Ukraine, then of the U.S.S.R.), had four reactors in place by 1986. With the planned addition of Units 5 and 6, for which foundation work was under way, the site would be the world's second largest electric power plant park, with an output of 6000 megawatts (electrical). The reactors were of a type called RBMK; they are graphite-moderated and use boiling-water pressure tubes.[22] Chernobyl and the Soviet nuclear power program were prominently featured in 1985 issues of the English-language periodical *Soviet Life*. The articles featured the safety of atomic energy and the low risk of accidents and radiation exposure. Meanwhile "Chernobyl" has become a household word because of a terrible reactor fire.

On April 25, 1986, a test was under way on Reactor 4 to determine how long the mechanical inertia of the turbine-generator's rotating mass could keep the generator turning and producing electric power after the steam supply was shut off (Figure 4–7). This was of interest because reactor coolant pumps and other

Figure 4–7

Schematic diagram of Reactor 4 at Chernobyl. This RBMK-type reactor produces steam for two 500-megawatt steam turbine generators, only one of which is shown. (*Adapted from John F. Ahearne, "Nuclear Power after Chernobyl,"* Science *236 (May 1987), p. 674.)*

Reprinted with permission from "Nuclear Power after Chernobyl," Science *236 (May 1987)* p. 674. Copyright 1987 American Association for the Advancement of Science.

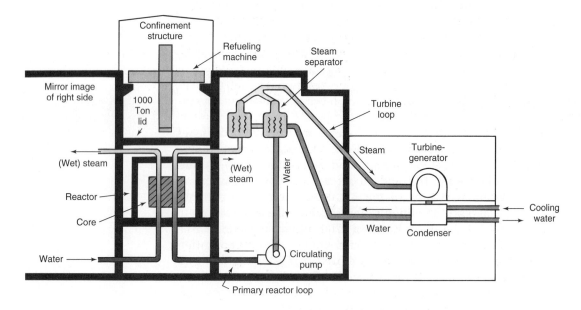

[22] David R. Marples, *Chernobyl and Nuclear Power in the USSR* (London: MacMillan Press, 1986); Mike Edwards, "Chernobyl—One Year After," *National Geographic* 171 (May 1987), pp. 632–53; Grigori Medvedev, *The Truth about Chernobyl* (New York: Basic Books, 1991); Alla Yaroshinskaya, *Chernobyl, the Forbidden Fruit* (Oxford: Jon Carpenter, 1994).

vital electric machinery must continue functioning though the generators may have had to be disconnected suddenly from a malfunctioning power grid. Special diesel generators will eventually start to provide emergency power for the plant, but diesel units cannot always be relied on to start promptly. This test was undertaken as part of a scheduled plant shutdown for general maintenance purposes.

It requires 3600 megawatts of thermal power in the RBMK reactor to produce 1200 megawatts at the generator output. Output of Unit 4 had been gradually throttled from 3200 megawatts (thermal) to 1600 megawatts and was to be slowly taken down to between 1000 and 700 megawatts, but at 2 P.M., the power dispatch controller at Kiev requested that output be maintained to satisfy an unexpected demand. This meant a postponement of the test. In preparation for the test, the reactor operators had disconnected the emergency core-cooling system so its power consumption would not affect the test results. This was to be the first of many safety violations.

Another error occurred when a control device was not properly reprogrammed to maintain power at the 700- to 1000-megawatt level. When at 11:10 P.M., the plant was authorized to reduce power, its output dropped all the way to 30 megawatts, where the reactor is difficult to control. Instead of shutting down the reactor, the operators tried to keep the test going by raising the control rods to increase power. Instead of leaving 15 controls inserted as required, the operators raised almost all control rods because, at the low power level, the fuel had become poisoned by a buildup of xenon-135, which absorbs neutrons.

The power output stayed steady at 200 megawatts (thermal)—still below what the test called for—but the test continued. In accordance with the test protocol, two additional circulating pumps were turned on to join the six already in operation. Under normal levels of power output, this would have contributed to the safety of the reactor, but at 200 megawatts, it required many manual adjustments to maintain the balance of steam and water. In addition, "the operators at this point recognized that because of the instabilities in this reactor and the way xenon poisoning builds up, once the reactor is shut down, they would have to wait a long time before starting it up again."[23] So, deciding to proceed with the test, the operators blocked the emergency signals and automatic shutdown controls because they would have been activated on removal of the electrical load.

[23] John F. Ahearne, "Nuclear Power after Chernobyl," *Science* 236 (May 1987), pp. 673–79; discussion by E. G. Silver and J. F. Ahearne, *Science* 238 (October 1987), pp. 144–45.

This left the reactor in a precarious position: "The reactor was now running free, isolated from the outside world, its control rods out, and its safety system disconnected." As Valery Legasov, then U.S.S.R. representative to the International Atomic Energy Commission (IAEC), reported at a review of the accident: "The reactor was free to do as it wished."[24]

At 1:23 A.M. the test began. When the steam valves were closed and its load was effectively removed, the reactor's power and temperature rose sharply. Unlike water-moderated reactors, the graphite-moderated RBMK reactor uses water only as a heat-transfer medium, not as a moderator. As the core becomes hotter, it allows fission to increase. This positive feedback effect produced a surge of power in Chernobyl's Reactor 4, from 7 percent to hundreds of times its rated thermal output: "The effect was the equivalent of $1/2$ ton of TNT exploding in the core. . . . The fuel did not have time to melt . . . it simply shattered into fragments."[25] The fuel, bereft of its cladding, came in contact with the water. A second explosion occurred (very likely a steam explosion). It lifted and shifted a 1000-ton concrete floor pad separating the reactor from the refueling area above it. The zirconium cladding of the fuel rods interacted with the circulating water to form hydrogen. This produced a spectacular display of fireworks. A shower of glowing graphite and fuel spewed over the compound while a radioactive plume was driven sky-high by the heat.

What followed was as inexcusable as what had caused the accident. While valiant firefighters lost their lives from radiation exposure while extinguishing the blaze, it took hours to warn the surrounding communities of radiation danger. Only when alert nuclear plant operators in faraway Sweden detected an increase in radioactivity did Moscow learn that something was amiss. The Soviet republics and the rest of Europe were not prepared to handle such a grave event, especially the radioactive fallout. Many countries blamed Moscow for not notifying them, but then neither did these countries have adequate monitoring systems of their own, not even to check on their local nuclear plants. Subsequent instructions on what to do about drinking milk, eating vegetables, letting children play outside, and other concerns of the populations of Europe depended more on the political leanings and the pronuclear or antinuclear stance of their respective health departments.

Acute radiation sickness, combined with burns, severely affected about 200 Chernobyl plant workers, of whom 31 died

---

[24] Nigel Hawkes, Geoffrey Lean, David Leigh, et al., *Chernobyl—The End of the Nuclear Dream* (New York: Vintage, 1986), p. 102.
[25] Ibid.

very quickly. Already by 1992, the number of excess death cases attributable to Chernobyl were reported to have exceeded 6000 within the Ukraine alone, and in Belarus, so many children have enlarged thyroids that 10,000 cancer cases above the usual can be expected.[26] The 1000 families living in a workers' settlement one mile from the plant were evacuated 12 hours after the explosion, but the plant had no responsibility for, nor direct link with, the communities beyond a 1.5-mile radius. The evacuation of nearby Pripyat and 71 villages within 18 miles of the plant started the next day. About 135,000 people had to be moved by buses and trucks. Numerous new villages were constructed to house the displaced. By arbitrarily announcing a new, "safe" radiation dose pegged at 10 times the former level, the Politbureau saved itself the burden of evacuating another 1.25 million people from surrounding areas and having to give medical care to countless more exposed to radiation. High government officials, however, were quietly evacuating their families from cities as far away as Kiev while the masses were asked to turn out for open-air May Day celebrations.

The near- and long-term effects of radiation on the people and fauna of Europe will be widely discussed for many years. Seven years after the accident, it had become clear that more radiation escaped than had been estimated earlier because the red glow that was targeted by the pilots with their loads of sand, clay, and dolomite came not from the reactor as thought but from a small, ejected core element 50 feet from the reactor. Contamination was also spread by the deliberate mixing of affected agricultural products with clean products that were then exported to other parts of the Soviet Union.

After the accident, the reactor was encased in a concrete sarcophagus, but not an airtight one—should the precariously perched reactor lid slide off, the ensuing radioactive dust could easily escape. But tunnels were dug underneath the reactor to install cooling pipes carrying liquid nitrogen. The tunnels also served to lay down a concrete layer to prevent leakage of radioactive water to the aquifer.

Strengthening a Skyscraper and Winning a Race against Time

The Citicorp tower case in New York City highlights the need for monitoring projects even after the participating engineers have gone on to other tasks. The occasion for reviewing the design of the steel structure arose in an unplanned way, but fortunately

---

[26] Testimony by Murray Feshbach before U.S. Senate Hearing in 1992; S.Hrg.102=765.

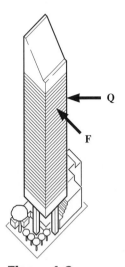

Figure 4–8
Axonometric view of
Citicorp tower with the
church in the lower left
hand corner. Wind loads:
F, frontal and Q,
Quartering. (*Adaptation of
an axonometric drawing
by Henry Dong, Anspach
Grossman Portugal, Inc.,
in Buildings Type Study
492,* Architectural Record,
*Mid-August Special Issue,
1976, p. 66.)*

construction problems were detected and resolved in a timely manner before an anticipated hurricane might have damaged the skyscraper.

Architect Hugh Stubbins and structural engineer Bill LeMessurier faced a big problem when they worked on the plans for New York's fifth highest skyscraper. St. Peter's Lutheran Church owned and occupied a corner of the lot designated in its entirety as the site for the new structure. An agreement was reached: the bank tower would rise from nine-story-high stilts positioned at the center of each side of the tower, and the church would be offered a brand new St. Peter's standing freely underneath one of the cantilevered corners. Completed in 1977, the Citcorp Center appears as shown in Figure 4–8. The new church building is seen below the lower left corner of the raised tower.

LeMessurier's structure departed from the usual—and therefore represented a real experiment—in that the massive stilts are not situated at the corners of the building, and half of its gravity load as well as all of its wind load are brought down an imaginatively designed trussed frame, which incorporates wind braces, on the outside of the tower.[27] In addition, LeMessurier installed a tuned mass damper, the first of its kind in a tall building, to keep the building from swaying in the wind.

As Joe Morgenstern relates in a *New Yorker* story,[28] questions asked by an engineering student a year after the tower's completion prompted LeMessurier to review certain structural aspects of the tower and pose some questions of his own. For instance, could the structure withstand certain loads due to strong quartering winds? In such cases, two sides of the building receive the oblique force of the wind, and the resultant force is 40 percent larger than when the wind hits only one face of the structure straight on. The only requirement stated in the building code specified adequacy to withstand certain perpendicular wind loads, and that was the basis for the design of the wind braces. It was determined there was no need to worry since the braces as designed could handle the excess oblique load without difficulty, provided the welds were of the expected high quality.

Nevertheless, the student's questions prompted LeMessurier to place a call from his Cambridge, Massachusetts, office to his

[27] Buildings Type Study 492, "Engineering for Architecture: Citicorp Center and St. Peter's Lutheran Church," *Architectural Record,* Mid-August Special Issue, 1976), pp. 61–71; Charles Thornton, "Conversation with William LeMessurier," *Exposed Structures in Building Design,* C. H. Thornton et al. (New York: McGraw-Hill, 1993).

[28] Joe Morgenstern, "The Fifty-Nine Story Crisis," *The New Yorker,* May 29, 1995, pp. 45–53. Check also the Online Ethics Center for Engineering and Science (http://ethics.cwru.edu/).

New York office, and he asked Stanley Goldstein, his engineer in charge of the tower erection, how the welded joints of the bracing structure had worked out. How difficult was the job? How good was the workmanship? To his dismay, Goldstein answered, "Oh, didn't you know? [The joints] were never welded at all because Bethlehem Steel came to us and said they didn't think we needed to do it." The New York office, as it was allowed to do, had approved the proposal that the joints be bolted instead. But again the diagonal winds had not been taken into account.

LeMessurier was not too concerned; after all, the tuned mass damper would still take care of the sway. So he turned to his consultant on the behavior of high buildings in wind, Alan Davenport at the University of Western Ontario. On reviewing the results of his earlier windtunnel tests on a scaled-down Citicorp Center, Davenport reported that a diagonal wind load would exceed the perpendicular wind load by much more than the 40 percent increase in stress predicted by an idealized mathematical model. Winds sufficient to cause failure of certain critical bolted joints—and therefore of the building—could occur in New York every 16 years.

Fortunately those braces that required strengthening were accessible, but the job would be disruptive and expensive, exceeding the insurance LeMessurier carried. What to do? As Morgenstern tells it, the eminent structural engineer retreated to his summerhouse on an island on Sebago Lake in Maine. There, in the quiet, he worked once more through all the design and wind tunnel numbers.

> [Then] he considered his options. Silence was one of them; only Davenport knew the full implications of what he had found, and he would not disclose them on his own. Suicide was another; if LeMessurier drove along the Maine Turnpike at a hundred miles an hour and steered into a bridge abutment, that would be that. But keeping silent required betting other people's lives against the odds, while suicide struck him as a coward's way out (and—although he was passionate about nineteenth-century classical music—unconvincingly melodramatic). The insight that seized him an instant later was entirely convincing, because it was so unexpected, giving him an almost giddy sense of power. "I had information that nobody else in the world had," LeMessurier recalls. "I had power in my hands to effect extraordinary events that only I could initiate. I mean, sixteen years to failure—that was very simple, very clearcut. I almost said, 'Thank you, dear Lord, for making this problem so sharply defined that there's no choice to make.'"[29]

---

[29] Ibid.

Now the direction to take was clear and LeMessurier acted quickly. He and Stubbins (the architect) met with their insurers, lawyers, the bank management, and the city building department to describe the problem. A retrofit plan was agreed upon: The wind braces would be strengthened at critical locations "by welding two-inch-thick steel plates over each of more than 200 bolted joints." Journalists, at first curious about the many lawyers converging on the various offices, disappeared when New York's major newspapers were shut down by a strike. The lawyers sought the advice of Leslie Robertson, a structural engineer with experience in disaster management. He alerted the mayor's Office of Emergency Management and the Red Cross so the surroundings of the building could be evacuated in case of a high wind alert. Robertson also arranged for a network of strain gauges to be attached to the structure at strategic points. This instrumentation allowed actual strains experienced by the steel to be monitored at a remote location. LeMessurier insisted on the installation of an emergency generator to assure uninterrupted availability of the damper.

When hurricane Ella appeared off the coast, there was some cause for worry, but work on the critical joints had almost been completed. Eventually the hurricane veered off and evacuation was not required. Even so, the retrofit and the tuned mass damper had been readied to withstand as much as a 200-year storm.

No litigation ever ensued. The parties were able to settle out of court, with Stubbins held blameless; LeMessurier and his joint-venture partners were charged the $2 million his insurance agreed to pay. The total repair bill had amounted to over $12.5 million.

As Morgenstern points out, the crisis produced no villains, only heroes. "LeMessurier emerged with his reputation not merely unscathed, but enhanced." We consider him a good example of an engineer who knew how to conduct an "experiment" down to the all-important monitoring of the final product, the tower.

Safe Exit

It is almost impossible to build a completely safe product or one that will never fail. The best one can do is to assure that when a product fails, (1) it will fail safely, (2) the product can be abandoned safely, or—at least—(3) the user can safely escape the product. Let us refer to these three conditions as *safe exit*. It is not obvious who should take the responsibility for providing safe exit. But apart from questions of who will build, install, maintain, and pay for a safe exit system, there remains the crucial question of who will recognize the need for a safe exit.

It is our position that providing for a safe exit is an integral part of the experimental procedure—in other words, of sound engineering. The experiment is to be carried out without causing bodily or financial harm. If safety is threatened, the experiment must be terminated. The full responsibility cannot fall on the shoulders of a lone engineer, but one can expect the engineer to issue warnings when a safe exit does not exist or the experiment must be terminated. The only way one can justify continuation of an experiment without safe exit is for all participants (including the subjects of the experiment) to have given valid consent for its continuation.

Here are some examples of what this might involve. Ships need lifeboats with sufficient spaces for all passengers and crew members. Buildings need usable fire escapes. Operation of nuclear power plants calls for realistic means of evacuating nearby communities. The foregoing are examples of safe exits for people. Provisions are also needed for safe disposal of dangerous products and materials: Altogether too many truck accidents and train derailments have exposed communities to toxic gases, and too many dumps have let toxic wastes get to the groundwater table or into the hands of children. Finally, to avoid system failure may require redundant or alternative means of continuing a process when the original process fails. Examples would be backup power systems for computer-based data banks, air traffic control systems, hospitals, and water systems.

Discussion Topics

1. Discuss what you see as the main similarities and differences in Three Mile Island and Chernobyl.

2. It has been said that Three Mile Island showed us the risks of nuclear power and the Arab oil embargo, the risk of having no energy. Forcing hazardous products or services from the market has been criticized as closing out the options of those individuals or countries with rising aspirations who can now afford them and who may all along have borne more than their share of the risks without any of the benefits. Finally, pioneers have always exposed themselves to risk. Without risk, there would be no progress. Discuss this problem of "the risk of no risk."[30]

3. Discuss the notion of safe exit, using evacuation plans for communities near nuclear power plants or chemical process plants. Refineries in the San Francisco Bay Area are currently required to jointly develop maps showing areas possibly exposed to dangerous air contamination resulting from refinery fires and explosions.

---

[30] Aaron Wildavsky, "No Risk Is the Highest Risk of All," in *Ethical Problems in Engineering,* 2nd ed., ed. Albert Flores (Troy, NY: Rensselaer Polytechnic Institute, 1980), pp. 221–26.

4. Valery Legasov, nuclear engineer and later U.S.S.R. representative to the IAEC, played an important role in containing the reactor fire at Chernobyl. He said that human errors brought on the accident, and if once we looked at safety technology as a means of protecting us from machines, now technology must be protected from us.[31] Discuss the change in Legasov's viewpoint and in what ways you agree or disagree with his statement.

5. Obtain and read a copy of each of the following articles having to do with safety concerns—or lack thereof—in the early history of atomic power: (a) Adolph J. Ackerman, "The Engineer's Responsibility and Duty to Dissent," and (b) John W. Foster, "An Example of an Engineer's Dissent," both published in *Consulting Engineer,* v.25, n.3, Sept. 1965, pp. 138–40 and 141–44 respectively. Then discuss the role of accident insurance as a possible self-regulating mechanism to assure serious attention to safety measures.

6. The toxic waste cases known as the "Love Canal" and "Woburn Cancer Cluster" episodes have received wide attention in the United States and are well documented in the technical and popular literature. The latter is the topic of a movie, *A Civil Action,* based on a book by Jonathan Harr in which he details lawyer Jan Schlichtmann's efforts to see justice done on behalf of some of Woburn's families. Compare these two toxic waste cases and discuss the bottlenecks that prevent release of pertinent information to public bodies and litigants in such court cases.

---

[31] Mike Moore, editorial, *Bulletin of Atomic Science,* May–June 1996.

5

Workplace Responsibilities and Rights

Data General Corporation grew spectacularly during its first decade of operation, quickly becoming a Fortune 500 company that was ranked third in overall sales of small computers. However, it began to fall behind the competition and desperately needed a powerful new microcomputer to sustain its share of the market. The development of that computer is chronicled by Tracy Kidder in his Pulitzer Prize–winning book *The Soul of a New Machine.*

Tom West, one of Data General's most trusted engineers, convinced management that he could build the new computer within one year—an unprecedented time for a project of its importance. West assembled a team of 15 exceptionally motivated though relatively inexperienced young engineers, many of whom were just out of school. Within six months, they designed the central processing unit, and they delivered the complete computer ahead of schedule. Named the Eclipse MV/8000, the computer immediately became a major marketing success.

The remarkable success was possible because the engineers came to identify themselves with the project and the product: "Ninety-eight percent of the thrill comes from knowing that the thing you designed works, and works almost the way you expected it would. If that happens, part of you is in that machine."[1] Yet, the "soul" of the new machine was not any one person. Instead, it was the team of engineers who invested themselves in the product through their personal commitment to work together creatively with colleagues as part of a design group. As might be expected, personality clashes occurred during the sometimes

---

[1] Tracy Kidder, *The Soul of a New Machine* (New York: Avon Books, 1981), p. 273.

143

frenzied work schedule, but conflicts were minimized by a commitment to teamwork, collegiality, shared commitment, and identification with the group's project.

This kind of performance understandably ranks high on the list of expectations that employers have of the engineers they employ or of the engineers they engage as consultants. Engineers in turn should see top performance at a professional level as their main responsibility, followed by others such as maintaining confidentiality and avoiding conflicts of interest. But engineers must also be given the opportunity to perform responsibly, and this means that their professional rights must be observed.

In this chapter we first address the needs to maintain confidentiality and to avoid conflicts of interest as issues related to an engineer's responsibilities as an employee. Next we discuss an engineer's rights as a professional and as an employee. We end with the issue of whistle-blowing that can arise when loyalty to the employer is in conflict with the engineer's personal ethics and the engineer's duty as a citizen.

Issues of Responsibility, Confidentiality, and Conflicts of Interest

Maintaining confidentiality and avoiding harmful conflicts of interest are especially important aspects of teamwork and trustworthiness. Let us begin with confidentiality and then turn to conflicts of interest, in each case seeking a clearer understanding of what is at stake morally.

Confidentiality: Definition

The duty of confidentiality is the duty to keep secret all information deemed desirable to keep secret. Deemed by whom? Basically, any information that the employer or client would like to have kept secret in order to compete effectively against business rivals. Often, this is understood to be any data concerning the company's business or technical processes that are not already public knowledge. While this criterion is somewhat vague, it clearly points to the employer or client as the main source of the decision as to what information is to be treated as confidential.

"Keep secret" is a relational expression. It always makes sense to ask, "Secret with respect to whom?" In the case of some government organizations, such as the FBI and CIA, highly elaborate systems for classifying information have been developed that identify which individuals and groups may have access to what information. Within other governmental agencies and private companies, engineers and other employees are usually expected to withhold information labeled "confidential" from unauthorized people both inside and outside the organization.

Several related terms should be distinguished. *Privileged information* literally means "available only on the basis of special privilege," such as the privilege accorded an employee working on a special assignment. *Proprietary information* is information that a company owns or is the proprietor of, and hence is a term carefully defined by property law. A rough synonym for "proprietary information" is *trade secret,* which can be virtually any type of information that has not become public, that an employer has taken steps to keep secret, and that is thereby given limited legal protection in common law (law generated by previous court rulings) that forbids employees from divulging it. *Patents* legally protect specific products from being manufactured and sold by competitors without the express permission of the patent holder. Trade secrets have no such protection, and a corporation may learn about a competitor's trade secrets through legal means—for instance, "reverse engineering," in which an unknown design or process can be traced out by analyzing the final product. But patents do have the drawback of being public and thus allowing competitors an easy means of working around them by finding alternative designs.

Confidentiality and Changing Jobs

The obligation to protect confidential information does not cease when employees change jobs. If it did, it would be impossible to protect such information. Former employees would quickly divulge it to their new employers, or perhaps for a price sell it to competitors of their former employers. Thus, the relationship of trust between employer and employee in regard to confidentiality continues beyond the formal period of employment. Unless the employer gives consent, former employees are barred indefinitely from revealing trade secrets. This provides a clear illustration of the way in which the professional integrity of engineers involves much more than mere loyalty to one's present employer.

Yet thorny problems arise in this area. Many engineers value professional advancement more than long-term ties with any one company and so change jobs frequently. Engineers in research and development are especially likely to have high rates of turnover. They are also the people most likely to be exposed to important new trade secrets. Moreover, when they transfer into new companies, they frequently do the same kind of work as before—precisely the type of situation in which trade secrets of their old companies may have relevance, a fact that could have strongly contributed to their having readily found new employment.

A high-profile case of trade secret violations in recent history was settled in January 1997 without coming to trial when VW (Volkswagen AG) agreed to pay GM (General Motors Corp. and GM's German subsidiary Adam Opel) $100 million in cash and to buy $1 billion in parts from GM over the next seven years. Why?

Because in March 1993, Jose Ignacio Lopez, GM's highly effective manufacturing expert, left GM to join VW (a fierce competitor in Europe) and took with him not only three colleagues and know-how, but also copies of confidential GM documents.

Instead of examining the Lopez case further, let us consider the legally significant case of Donald Wohlgemuth, a chemical engineer who at one time was manager of B.F. Goodrich's space suit division.[2] Technology for space suits was undergoing rapid development, with several companies competing for government contracts. Dissatisfied with his salary and the research facilities at B. F. Goodrich, Wohlgemuth negotiated a new job with International Latex Corporation as manager of engineering for industrial products. International Latex had just received a large government subcontract for developing the Apollo astronauts' space suits, and that was one of the programs Wohlgemuth would manage.

The confidentiality obligation required that Wohlgemuth not reveal any trade secrets of Goodrich to his new employer. But this was easier said than done. Of course, it is possible for employees in his situation to refrain from explicitly stating processes, formulas, and material specifications. Yet in exercising their general skills and knowledge, it is virtually inevitable that some unintended "leaks" will occur. An engineer's knowledge base generates an intuitive sense of what designs will or will not work, and trade secrets form part of this knowledge base. To fully protect the secrets of an old employer on a new job would thus virtually require that part of the engineer's brain be removed.

Is it perhaps unethical, then, for employees to change jobs in cases where unintentional revelations of confidential information are a possibility? Some companies have contended that it is. Goodrich, for example, charged Wohlgemuth with being unethical in taking the job with International Latex. Goodrich also went to court seeking a restraining order to prevent him from working for International Latex or any other company that developed space suits. The Ohio Court of Appeals refused to issue such an order, although it did issue an injunction prohibiting Wohlgemuth from revealing any Goodrich trade secrets. Their reasoning was that while Goodrich had a right to have trade secrets kept confidential, it had to be balanced against Wohlgemuth's personal right to seek career advancement. And this would seem to be the correct moral verdict as well.

---

[2] Michael S. Baram, "Trade Secrets: What Price Loyalty?" *Harvard Business Review,* November–December 1968. Reprinted in Deborah G. Johnson, ed., *Ethical Issues in Engineering* (Englewood Cliffs, NJ: Prentice Hall, 1991), pp. 279–90.

Confidentiality and Management Policies

What might be done to recognize the legitimate personal interests and rights of engineers and other employees while also recognizing the rights of employers in this area?[3] One approach is to use employment contracts that place special restrictions on future employment. Traditionally, those restrictions have centered on geographical location of future employers, length of time after leaving the present employer before one can engage in certain kinds of work, and the type of work it is permissible to do for future employers. Thus, Goodrich might have required as a condition of employment that Wohlgemuth sign an agreement that if he sought work elsewhere, he would not work on space suit projects for a competitor in the United States for five years after leaving Goodrich.

Yet such contracts are hardly agreements between equals, and they threaten the right of individuals to pursue their careers freely. For this reason, the courts have tended not to recognize such contracts as binding, although they do uphold contractual agreements forbidding disclosure of trade secrets.

A different type of employment contract is perhaps not so threatening to employee rights in that it offers positive benefits in exchange for the restrictions it places on future employment. Consider a company that normally does not have a portable pension plan. It might offer such a plan to an engineer in exchange for an agreement not to work for a competitor on certain kinds of projects for a certain number of years after leaving the company. Or another clause might offer an employee a special post-employment annual consulting fee for several years on the condition that he or she not work for a direct competitor during that period.

Other tactics aside from employment contract provisions have been attempted by various companies. One is to place tighter controls on the internal flow of information by restricting access to trade secrets except where absolutely essential. The drawback to this approach is that it may create an atmosphere of distrust in the workplace. It might also stifle creativity by lessening the knowledge base of engineers involved in research and development.

One potential solution is for employers to help generate a sense of professional responsibility among their staff that reaches beyond merely obeying the directives of current employers. Engineers can then develop a real sensitivity to the moral conflicts to which they may be exposed by making certain job

---

[3] Ibid., pp. 285–90.

changes within the company. They can arrive at a greater appreciation of why trade secrets are important in a competitive system and learn to take the steps necessary to protect them. In this way, professional concerns and employee loyalty can become intertwined and reinforce each other.

Confidentiality: Justification

Upon what moral basis does the confidentiality obligation rest, with its wide scope and obvious importance? The primary justification is to respect the autonomy (freedom, self-determination) of individuals and corporations and to recognize their legitimate control over some private information concerning themselves.[4] Without that control, they could not maintain their privacy and protect their self-interest insofar as it involves privacy. Just as patients should be allowed to maintain substantial control over personal information, so employers should have some control over the private information about their companies. All the major ethical theories recognize the importance of autonomy, whether it is understood in terms of rights to autonomy, duties to respect autonomy, the utility of protecting autonomy, or the virtue of respect for others.

Additional justifications include trustworthiness: Once practices of maintaining confidentiality are established socially, trust and trustworthiness can grow. Thus, when clients go to attorneys or tax accountants, they expect them to maintain confidentiality, and the professional indicates that confidentially will be maintained. Similarly, employees often make promises (in the form of signing contracts) not to divulge certain information considered sensitive by the employer.

In addition, there are public benefits in recognizing confidentiality relationships within professional contexts. For example, if patients are to have the best chances of being cured, they must feel completely free to reveal the most personal information about themselves to physicians, and that requires trust that the physician will not divulge private information. Likewise, the economic benefits of competitiveness within a free market are promoted when companies can maintain some degree of confidentiality concerning their products. Developing new products often requires investing large resources in acquiring new knowledge. The motivation to make those investments might diminish if that knowledge were immediately dispersed to competitors who could then quickly make better products at lesser cost, since they did not have to make comparable investments in research and development.

---

[4] Sissela Bok, *Secrets* (New York: Pantheon Books, 1982), pp. 116–35.

It must not be overlooked that confidentiality has its limits, particularly when it is invoked to hide misdeeds. Investigations into a wide variety of white collar crimes covered up by management in industry or public agencies have been thwarted by invoking confidentiality or false claims of secrecy based on national interest.

Conflicts of Interest: Definition and Examples

We turn now to some equally thorny issues concerning conflicts of interest. Professional conflicts of interest are situations where professionals have an interest that, if pursued, might keep them from meeting their obligations to their employers or clients. Sometimes such an interest involves serving in some other professional role, say, as a consultant for a competitor's company. Other times it is a more personal interest, such as making substantial private investments in a competitor's company.

Concern about conflicts of interest largely centers on their potential to distort good judgment in faithfully serving an employer or client.[5] Exercising good judgment means arriving at beliefs on the basis of expertise and experience, as opposed to merely following simple rules. Thus, we can refine our definition of conflicts of interest by saying that they typically arise when *two* conditions are met: (1) the professional is in a relationship or role that requires exercising good judgment on behalf of the interests of an employer or client and (2) the professional has some additional or side interest that could threaten good judgment in serving the interests of the employer or client—either the good judgment of that professional or the judgment of a typical professional in that situation. Why the reference to "a typical professional"? There might be conclusive evidence that the actual professionals involved would never allow a side interest to affect their judgment, yet they are still held to be in a conflict of interest.

"Conflict of interest" and "conflicting interests" are not synonyms.[6] A student, for example, may have interests in excelling on four final exams. She knows, however, that there is time to study adequately for only three of them, and so she must choose which interest not to pursue. In this case, "conflicting interests" means a person has two or more desires that cannot all be satisfied

---

[5] Michael Davis, "Conflict of Interest," *Business and Professional Ethics Journal* 1 (Summer 1982), pp. 17–27; Paula Wells, Hardy Jones, and Michael Davis, *Conflicts of Interest in Engineering* (Dubuque, IA: Kendall/Hunt, 1986).

[6] Joseph Margolis, "Conflict of Interest and Conflicting Interests," in *Ethical Theory and Business,* ed. T. Beauchamp and N. Bowie (Englewood Cliffs, NJ: Prentice Hall, 1979), p. 361.

given the circumstances. But there is no suggestion that it is morally wrong to try pursuing them all. By contrast, in professional conflicts of interest, it is often physically or economically possible to pursue all of the conflicting interests, but doing so could be morally problematic.

Because of the great variety of possible outside interests, conflicts of interest can arise in innumerable ways, and with many degrees of subtlety. We will sample only a few of the more common situations.

Gifts, Bribes, and Kickbacks

A bribe is a substantial amount of money or goods offered beyond a stated business contract with the aim of winning an advantage in gaining or keeping the contract. "Substantial" is a vague term, but it alludes to amounts, beyond acceptable gratuities, that are sufficient to distort the judgment of a typical person. Typically, though not always, bribes are made in secret. Gifts are not bribes as long as they are small gratuities offered in the normal conduct of business, where a small gift used to be jokingly referred to as having the value of something you could "eat, drink, or smoke in a day." Prearranged payments made by contractors to companies or their representatives in exchange for contracts actually granted are called "kickbacks." When suggested by the granting party to the party bidding on the contract, the latter often defends its participation in such an arrangement as having been subjected to "extortion."

> A gift is not a bribe if you can eat, drink, or smoke it in a day.
>
> —Old-timers' saying

Often, companies give gifts to selected employees of government agencies or partners in trade. Many such gifts are unobjectionable, some are intended as bribes, and still others create conflicts of interest that do not, strictly speaking, involve bribes. What are the differences? In theory, these distinctions may seem clear, but in practice they become blurry. Bribes are illegal or immoral because they are substantial enough to threaten fairness in competitive situations, while gratuities are of smaller amounts. Some gratuities play a legitimate role in the normal conduct of business, while others can bias judgment like a bribe does. Much depends on the context, and there are numerous gray areas, which is why companies often develop elaborate guidelines for their employees.

Texas Instruments, for example, not only discusses gifts in its policy manual but it also makes available detailed brochures illustrating acceptable and unacceptable gifts within particular contexts. One context is doing business with the U.S. Department of Defense, whose officials are prohibited by federal law from accepting "anything of value" that bears in any way on official government business. Does that mean a Texas Instrument official cannot buy a moderately priced dinner for a Department of Defense official who is visiting? Yes. Does it mean the Texas Instruments employee cannot take the government official to the airport in the company's courtesy van? Maybe. Transportation to the airport may have a substantial value, as does company time. Texas Instruments allows its employees to offer the government official a ride to the airport if "1. The use of commercial transportation is impractical, and 2. Refusing your offer would interfere significantly with their [i.e., the officials'] performance of official duties."[7] Such serpentine policies may seem silly, but in fact they represent good-faith efforts to avoid even the appearance of a conflict of interest in situations of intense public scrutiny and cynicism.

What about more routine business contexts? Is it all right to accept the occasional luncheon paid for by vendors giving sales presentations, or a gift one believes is given in friendship rather than for influence? The guidelines for use with the principles of ethics of ASCE (Sec. 4c) or ASME (Sec. 4e) recommend a hard line on such gratuities: "Engineers shall not solicit nor accept gratuities, directly or indirectly, from contractors, their agents, or other parties dealing with their clients or employers in connection with work for which they are responsible." Many employers would consider this recommendation extreme. Company policies generally ban any gratuities that have more than nominal value or exceed widely and openly that accepted as normal business practice. An additional rule of thumb is: "If the offer or acceptance of a particular gift could have embarrassing consequences for your company if made public, then do not accept the gift."

Interests in Other Companies

Some conflicts of interest consist in having an interest in a competitor's or a subcontractor's business. One blatant example is actually working for the competitor or subcontractor as an employee or consultant. Another example is partial ownership or substantial stockholdings in the competitor's business. Does

---

[7] *Cornerstone* 2, TI Ethics Office, Texas Instruments, Richardson, Texas, 1989, p. 2.

holding a few shares of stock in a company with which one has occasional dealings constitute a conflict of interest? Usually not, but as the number of shares of stock increases, the issue becomes blurry. Again, is there a conflict of interest if one's spouse works for a subcontractor to one's company? Usually not, but a conflict of interest arises if one's job involves granting contracts to that subcontractor.

Should there be a general prohibition on *moonlighting,* that is, working in one's spare time for another company? That would violate the rights to pursue one's legitimate self-interest. Moonlighting usually creates conflicts of interest only in special circumstances, such as working for competitors, suppliers, or customers. Even then, in rare situations, an employer sometimes gives permission for exceptions, for example, when the experience gained would greatly promote business interests. A special kind of conflict of interest arises, however, when moonlighting leaves one exhausted and thereby harms job performance.[8]

Insider Information

An especially sensitive conflict of interest consists in using "inside" information to gain an advantage or set up a business opportunity for oneself, one's family, or one's friends. The information might concern one's own company or another company with which one does business. For example, engineers might tell their friends about the impending announcement of a revolutionary invention their company has been perfecting, or of their company's plans for a merger that will greatly improve the worth of another company's stock. In doing so, they give those friends an edge on an investment promising high returns. Owning stock in the company for which one works is of course not objectionable, and this is often encouraged by employers. But such ownership should be based on the same information available to the general public.

While conflict of interest occurs in many professional areas, it may come as a surprise that it also occurs in academic settings. Professors are frequently hired as consultants by companies that need the academic's expertise, and many professors in the sciences and engineering establish their own companies or form partnerships to develop commercially what they have learned through their research at the university. Here are two examples of the problems that can arise.

---

[8] George L. Reed, "Moonlighting and Professional Responsibility," *Journal of Professional Activities: Proceedings of the American Society of Civil Engineers* 96 (September 1970); pp. 19–23.

Experts at public as well as private universities turned down requests by the office of the Attorney General of California to testify for the state in its damage suit against Union Oil and others who were drilling offshore at Santa Barbara when the massive oil leak occurred in 1969. State officials believe that the experts (among them engineering professors) refused to offer expert testimony because they were afraid of losing industry grants and consulting arrangements if they were to testify unfavorably in the eyes of the oil industry.

A professor of electrical engineering at a west coast university was found to have used $144,000 in grant funds to purchase electronic equipment from a company he owned in part. He had not revealed his ownership to the university; he had priced the equipment much higher than market value, and some of the purchased items were never received. The Supplier Information Form and Sole Source Justification Statements had been submitted as required, but with falsified content. In addition, the professor had hired a brother and two sisters for several years, concealing their relationship to him in violation of anti-nepotism rules and paying them for research work they did not perform. All told he had defrauded the university of at least $500,000 in research funds. Needless to say, the professor lost his university position and had to stand trial in civil court when an internal audit and subsequent hearings revealed these irregularities.

These cases indicate that conflicts of interest are not restricted to industry. The academic biotechnology field, with its many professor-entrepreneurs, is particularly vulnerable in this regard.

Moral Status of Conflicts of Interest

What is wrong with employees having conflicts of interest? Most of the answer is obvious from our stated definition: Employee conflicts of interest occur when employees have interests that if pursued could keep them from meeting their obligations to serve the interests of the employer or client for whom they work. Such conflicts of interest should be avoided because they threaten to prevent one from fully meeting those obligations.

More than this needs to be said, however. Why should mere threats of possible harm always be condemned? Suppose that substantial good might sometimes result from pursuing a conflict of interest?

In fact, it is not always unethical to pursue conflicts of interest. In practice, some conflicts are thought to be unavoidable, or even acceptable. One illustration of this is that the government allows employees of aircraft manufacturers to serve as government inspectors for the Federal Aviation Agency (FAA). The FAA is charged with regulating airplane manufacturers and making objective safety and quality inspections of the airplanes they build. Naturally, the two roles of (1) government inspector and (2) employee of the manufacturer being inspected could bias judgments. Yet with careful screening of inspectors, the likelihood of such bias is said to be outweighed by the practical necessities of airplane inspection. The options would be to greatly increase the number of nonindustry government workers (at great expense to taxpayers) or to do without government inspection altogether (putting public safety at risk).

Even when conflicts of interest are unavoidable or reasonable, employees are still obligated to inform their employers and obtain approval. This suggests a fuller answer to why conflicts of interest are generally prohibited: (1) The professional obligation to employers is very important in that it overrides in the vast majority of cases any appeal to self-interest on the job and (2) the professional obligation to employers is easily threatened by self-interest (given human nature) in a way that warrants especially strong safeguards to ensure that it is fulfilled by employees.

Many conflicts of interest violate trust, in addition to undermining specific obligations. Employed professionals are in fiduciary (trust) relationships with their employers and clients. Allowing side interests to distort one's judgment violates that trust. And additional types of harm can arise as well. Many conflicts of interest are especially objectionable in business affairs precisely because they pose risks to free competition. In particular, bribes and large gifts are objectionable because they lead to awarding contracts for reasons other than the best work for the best price.

As a final point, we should note that even the appearance of conflicts of interest, especially appearances of seeking a personal profit at the expense of one's employer, is considered unethical since the appearance of wrongdoing can harm a corporation as much as any actual bias that might result from such practices.

Discussion Topics

1. Consider the following example:

   Who Owns Your Knowledge? Ken is a process engineer for Stardust Chemical Corp., and he has signed a secrecy agreement with the firm that prohibits his divulging information that the company considers proprietary.

   Stardust has developed an adaptation of a standard piece of equipment that makes it highly efficient for cooling a viscous plastics slurry. (Stardust decides not to patent the idea but to keep it as a trade secret.) Eventually, Ken leaves Stardust and goes to work for a candy-processing company that is not in any way in competition. He soon realizes that a modification similar to Stardust's trade secret could be applied to a different machine used for cooling fudge, and at once has the change made.[9]

Has Ken acted unethically?

2. American Potash and Chemical Corporation advertised for a chemical engineer having industrial experience with titanium oxide. It succeeded in hiring an engineer who had formerly supervised E. I. Du Pont de Nemours and Company's production of titanium oxide. Du Pont went to court and succeeded in obtaining an injunction prohibiting the engineer from working on American Potash's titanium oxide projects. The reason given for the injunction was that it would be inevitable that the engineer would disclose some of du Pont's trade secrets.[10] Defend your view as to whether the court injunction was morally warranted or not.

3. Consider the following case:

    Facts: Engineer Doe is employed on a full-time basis by a radio broadcast equipment manufacturer as a sales representative. In addition, Doe performs consulting engineering services to organizations in the radio broadcast field, including analysis of their technical problems and, when required, recommendation of certain radio broadcast equipment as may be needed. Doe's engineering reports to his clients are prepared in form for filing with the appropriate governmental body having jurisdiction over radio broadcast facilities. In some cases Doe's engineering reports recommend the use of broadcast equipment manufactured by his employer.

    Question: May Doe ethically provide consulting services as described?[11]

4. Consider the following case:

    Scott Bennett is the engineer assigned to deal with vendors who supply needed parts to the Upscale Company. Larry Newman, sales representative from one of Upscale's regular vendors, plays in the same golf league as Scott. One evening they go off in the same foursome. Sometime during the round Scott mentions

---

[9] Philip M. Kohn and Roy V. Hughson, "Perplexing Problems in Engineering Ethics," *Chemical Engineering* 87 (May 5, 1980), p. 102. Quotations in text used with permission of McGraw-Hill Book Co.

[10] Charles M. Carter, "Trade Secrets and the Technical Man," *IEEE Spectrum* 6 (February 1969), p. 54.

[11] *NSPE Opinions of the Board of Ethical Review,* Case 75.10, National Society of Professional Engineers, Washington, DC, web site: www.nspe.org/eh-home.asp.

that he is really looking forward to vacationing in Florida next month. Larry says his uncle owns a condo in Florida that he rents out during the months he and his family are up north. Larry offers to see if the condo is available next month—assuring Scott that the rental cost would be quite moderate.

What should Scott say?[12]

Does your answer turn on whether Scott's company policy indicates a clear answer to this question?

5. Junior colleges and even universities increasingly hire part-time lecturers to teach classes when they lack the funds to employ tenure-truck faculty, or when they think they can get more for less. But they usually will not pay a lecturer enough per course, nor engage any one lecturer for enough courses at a time, to allow a decent living. Nor will such part-time lecturers receive the usual health and retirement benefits. As a result, many lecturers become "freeway flyers" as they rush from one college to another to add to their meager income from each. Schools of engineering may hire a lecturer on a temporary basis when such a person can give instruction in a specialty field not covered by the regular faculty, or while a professorship is vacant. What do you see as advantages and disadvantages of hiring part-time lecturers? Are there dangers of *conflicting interests* when a lecturer's time is taken up by too many commitments at the same time? Could there also be *conflicts of interest?* Do these situations involve moonlighting?

Rights of Engineers

Engineers have several types of moral rights, which fall into the sometimes overlapping categories of human, employee, contractual, and professional rights. As *humans,* engineers have fundamental rights to live and freely pursue their legitimate interests, which implies, for example, rights not to be unfairly discriminated against in employment on the basis of sex, race, or age. As *employees,* engineers have special rights, including the right to receive one's salary in return for performing one's duties and the right to engage in the nonwork political activities of one's choosing without reprisal or coercion from employers. As *professionals,* engineers have special rights that arise from their professional role and the obligations it involves. We begin with professional rights, most of which can be viewed as aspects of a fundamental right of professional conscience. We now move to a discussion of professional rights, followed by employee rights.

---

[12] "The Condo," in *Teaching Engineering Ethics: A Case Study Approach,* ed. Michael S. Pritchard (Kalamazoo: Center for the Study of Ethics in Society, Western Michigan University, 1993).

Professional Rights

Basic Right of Professional Conscience

The right of professional conscience is the moral right to exercise professional judgment in pursuing professional responsibilities. Pursuing those responsibilities involves exercising both technical judgment and reasoned moral convictions. This right has limits, of course, and must be balanced against responsibilities to employers and colleagues of the sort discussed earlier.

If the duties of engineers were so clear that it was obvious to every sane person what was morally proper in every situation, there would be little point in speaking of "conscience" in specifying this basic right. Instead, we could simply say it is the right to do what everyone agrees is obligatory for the professional engineer to do. But engineering, like other professions, calls for morally complex decisions. It requires autonomous moral judgment in attempting to uncover the most morally reasonable courses of action, and the correct courses of action are not always obvious.

As with most moral rights, the basic professional right is an entitlement giving one the moral authority to act without interference from others. It is a "liberty right" that places an obligation on others not to interfere with its proper exercise. Yet, occasionally, special resources may be required by engineers seeking to exercise the right of professional conscience in the course of meeting their professional obligations. For example, conducting an adequate safety inspection may require that special equipment be made available by employers. Or, more generally, in order to feel comfortable about making certain kinds of decisions on a project, the engineers involved may need an environment conducive to trust and support, which management may be obligated to help create and sustain. In this way, the basic right is also in some respects a "positive right" placing on others an obligation to do more than merely not interfere.

> Character counts. Ethics is not for whimps.
> —Michael Josephson

The right of professional conscience implies more specific rights, corresponding to specific professional obligations. In the next part of this chapter, we discuss the right to whistle blow in some situations where the public good is severely threatened. Here we cite two further examples: the right of conscientious refusal and the right to recognition.

Right of Conscientious Refusal

The right of conscientious refusal is the right to refuse to engage in unethical behavior, and to refuse to do so solely because one views it as unethical. This is a kind of second-order right. It arises because other rights to honor moral obligations within the authority-based relationships of employment sometimes come into conflict.

There are two situations to be considered: (1) where there is widely shared agreement in the profession as to whether an act is unethical and (2) where there is room for disagreement among reasonable people over whether an act is unethical.

It seems clear enough that engineers and other professionals have a moral right to refuse to participate in activities that are illegal and uncontroversially unethical (for example, forging documents, altering test results, lying, giving or taking bribes, or padding payrolls). And coercing employees into acting by means of threats (to their jobs) plainly constitutes a violation of this right of theirs.

The troublesome cases concern situations where there is no shared agreement about whether a project or procedure is unethical. Do engineers have any rights to exercise their personal consciences in these more cloudy areas? Just as pro-life physicians and nurses have a right not to participate in abortions, engineers should be recognized as having a *limited* right to turn down assignments that violate their personal consciences in matters of great importance, such as threats to human life, even where there is room for moral disagreement among reasonable people about the situation in question. We emphasize the word "limited" because the right is contingent on the organization's ability to reassign the engineer to alternative projects without serious economic hardship to itself. The right of professional conscience does not extend to the right to be paid for not working.

Right to Recognition

Engineers have a right to professional recognition for their work and accomplishments. Part of this involves fair monetary remuneration and part, nonmonetary forms of recognition. The right to recognition, and especially fair remuneration, may seem to be purely a matter of self-interest rather than morality, but it is both. Without a fair remuneration, engineers cannot concentrate their energies where they properly belong—on carrying out the immediate duties of their jobs and on maintaining up-to-date skills through formal and informal continuing education. Their time will be taken up by money worries, or even by moonlighting in order to maintain a decent standard of living.

The right to reasonable remuneration is clear enough to serve as a moral basis for arguments against corporations that make excessive profits while their employees are poorly paid. It can also serve as the basis for criticizing the unfairness of patent arrangements that fail to give more than nominal rewards to the creative engineers who make the discoveries leading to the patents. If a patent leads to millions of dollars of revenue for a company, it is unfair to give the discoverer no more than a nominal bonus and a thank-you letter.

But the right to professional recognition is not sufficiently precise to pinpoint just what a reasonable salary is or what a fair remuneration for patent discoveries is. Such detailed matters must be worked out cooperatively between employers and employees, for they depend on both the resources of a company and the bargaining position of engineers. Professional societies can be of help by providing general guidelines.

Foundation of Professional Rights

There are two general ways to justify the basic right of professional conscience. One is to proceed piecemeal by reiterating the justifications given for the specific professional *duties*. Whatever justification there is for the specific duties will also provide justification for allowing engineers the *right* to pursue those duties. Fulfilling duties, in turn, requires the exercise of moral reflection and conscience, rather than rote application of simplistic rules. Hence, the justification of each duty ultimately yields a justification of the right of conscience with respect to that duty.

The second way is to justify the right of professional conscience, which involves grounding it more directly in the ethical theories. Thus, duty ethics regards professional rights as implied by general duties to respect persons, and rule-utilitarianism would accent the public good of allowing engineers to pursue their professional duties. Rights ethics would justify the right of professional conscience by reference to the rights of the public not to be harmed and the rights to be warned of dangers from the "social experiments" of technological innovation.

Employee Rights

Employee rights are any rights, moral or legal, that involve the status of being an employee. They overlap with some professional rights, of the sort just discussed, and they also include institutional rights created by organizational policies or employment agreements, such as the right to be paid the salary specified in one's contract or employment agreement. However, here we will focus on human rights that exist even if unrecognized by specific contract arrangements.

Many of these human rights are discussed more fully in *Freedom Inside the Organization* by David Ewing[13] (for many years

---

[13] David W. Ewing, *Freedom Inside the Organization* (New York: McGraw-Hill, 1977), pp. 234–35.

editor of *The Harvard Business Review,* 1949–85). Ewing refers to employee rights as the "black hole in American rights." The Bill of Rights in the Constitution was written to apply to government, not to business. But when the Constitution was written, no one envisaged the giant corporations that have emerged in our century. Corporations wield enormous power politically and socially, often in multinational settings, and they operate much as mini-governments and are often comparable in size to those governments the authors of the Constitution had in mind. For instance, American Telephone & Telegraph in the 1970s employed twice the number of people inhabiting the largest of the original 13 colonies when the Constitution was written.

Ewing proposes that large corporations ought to recognize a basic set of employee rights. As examples, we will discuss rights to privacy and to equal opportunity.

Privacy

The right to pursue outside activities can be thought of as a right to personal privacy in the sense that it means the right to have a private life off the job. In speaking of the right to privacy here, however, we mean the right to control access to and use of information about oneself. As with the right to outside activities, this right is limited in certain instances by employers' rights, but even then who among employers has access to confidential information is restricted. For example, the personnel division needs medical and life insurance information about employees, but immediate supervisors usually do not.

Consider a few examples of situations in which the functions of employers conflict with the right employees have to privacy:

1. Job applicants at the sales division of an electronics firm are required to take personality tests that include personal questions about alcohol use and sexual conduct. The rationale given for asking those questions is a sociological study showing correlations between sales ability and certain data obtained from answers to the questionnaire. (That study has been criticized by other sociologists.)

2. A supervisor unlocks and searches the desk of an engineer who is away on vacation without the permission of that engineer. The supervisor suspects the engineer of having leaked information about company plans to a competitor and is searching for evidence to prove those suspicions.

3. A large manufacturer of expensive pocket computers has suffered substantial losses from employee theft. It is believed that more than one employee is involved. Without notifying employees, hidden surveillance cameras are installed.

4. A rubber products firm has successfully resisted various attempts by a union to organize its workers. It is always one step ahead of the union's strategies, in part because it monitors the phone calls of employees who are union sympathizers. It also pays selected employees bonuses in exchange for their attending union meetings and reporting on information gathered. It considered, but rejected as imprudent, the possibility of bugging the rest areas where employees were likely to discuss proposals made by union organizers.

We may disagree about which of these examples involve abuse of employer prerogatives. Yet the examples remind us of both the importance of privacy (as discussed in Chapter 2) and how easily rights of privacy are abused. Employers should be viewed as having the same trust relationship with their employees concerning confidentiality that doctors have with their patients and lawyers have with their clients.[14] In all of these cases, personal information is given in trust on the basis of a special professional relationship.

Equal Opportunity: Nondiscrimination

Perhaps nothing is more demeaning than to be discounted because of personal attributes such as one's race, skin color, age, politics, or religious outlook. These aspects of biological makeup and basic conviction lie at the heart of self-identity and self-respect. Such discrimination—that is, morally unjustified treatment of people on arbitrary or irrelevant grounds—is especially pernicious within the work environment, for work is itself fundamental to a person's self-image. Accordingly, human rights to fair and decent treatment at the workplace and in job training are vitally important.

Consider the following examples:

1. An opening arises for a chemical plant manager. Normally such positions are filled by promotions from within the plant. The best qualified person in terms of training and years of experience is an African-American engineer. Management believes, however, that the majority of workers in the plant would be disgruntled by the appointment of a nonwhite manager. They fear lessened employee cooperation and efficiency. They decide to promote and transfer a white engineer from another plant to fill the position.

---

[14] Mordechai Mironi, "The Confidentiality of Personnel Records," *Labor Law Journal* 25 (May 1974), p. 289.

2. A farm equipment manufacturer has been hit hard by lowered sales caused by a flagging produce economy. Layoffs are inevitable. During several clandestine management meetings, it is decided to use the occasion to "weed out" some of the engineers within 10 years of retirement in order to avoid payments of unvested pension funds.

These examples involve discrimination. They also involve violation of antidiscrimination laws, in particular the Civil Rights Act of 1964: "It shall be an unlawful employment practice for an employer to fail or refuse to hire or to discharge any individual, or otherwise to discriminate against any individual with respect to his compensation, terms, conditions, or privileges of employment, because of such individual's race, color, religion, sex, or national origin" (Title VII, Equal Employment Opportunity). Age discrimination was added in the 1967 Age Discrimination in Employment Act, and discrimination based on disability was forbidden in the 1994 Americans With Disabilities Act.

Equal Opportunity: Sexual Harassment

Beginning in 1991, several events focused national attention on sexual harassment. In October of that year, Anita Hill testified against confirming Supreme Court nominee Clarence Thomas, charging that he made lewd remarks and unwanted sexual provocations to her years earlier when she had worked for him at the Justice Department. Hill was a respected attorney and law professor, and at the time, one-third of Americans were convinced she was telling the truth. The majority of the Senate Hearing Committee sided with Clarence Thomas, however, and he was confirmed as Supreme Court justice amid controversies over what sexual harassment is, how it is to be proven, and how best to prevent it. A series of scandals followed in the military, first the Navy and then the Army, and quickly corporations were caught up in a fundamental social debate about what sexual harassment is. More recently officers of the highest ranks in the U.S. military have been discharged for engaging in sexual relations with wives of subordinates, even when conducted on a consensual basis. President Clinton did not escape censure when found to have been involved in an unseemly sexual liaison with a White House intern, not against her will. Persons in high places must recognize that the aura of importance can attract members of the opposite sex and that exploiting this tendency can damage the working climate and the real business that should be conducted at the workplace.

One definition of sexual harassment is "the unwanted imposition of sexual requirements in the context of a relationship of unequal power."[15] It takes two main forms: *quid pro quo* and hostile work environment.

Quid pro quo includes cases where supervisors require sexual favors as a condition for some employment benefit (a job, promotion, or raise). It can take the form of a sexual threat (of harm) or sexual offer (of a benefit in return for a benefit). *Hostile work environment,* by contrast, is any sexually oriented aspect of the workplace that threatens employees' rights to equal opportunity. It includes unwanted sexual proposals, lewd remarks, sexual leering, posting of nude photos, and inappropriate physical contact.

What is morally objectionable about sexual harassment? Sexual harassment is a particularly invidious form of sex discrimination, involving as it does not only the abuse of gender roles and authority relationships, but the abuse of sexual intimacy itself. Sexual harassment is a display of power and aggression through sexual means. Accordingly, it has appropriately been called "dominance eroticized."[16] Insofar as it involves coercion, sexual harassment constitutes an infringement of one's autonomy to make free decisions concerning one's body. But whether or not coercion and manipulation are used, it is an assault on the victim's dignity. In abusing sexuality, such harassment degrades people on the basis of a biological and social trait central to their sense of personhood.

Thus, a duty ethicist like Kant would condemn it as violating the duty to treat people with respect, to treat them as having dignity and not merely as means to personal aggrandizement and gratification of one's sexual and power interests. A rights ethicist would see it as a serious violation of the human right to pursue one's work free from the pressures, fears, penalties, and insults that typically accompany sexual harassment. And a utilitarian would emphasize the impact it has on the victim's happiness and self-fulfillment, and on women in general. This also applies to men who experience sexual harassment.

Equal Opportunity: Preferential Treatment

Preferential treatment, as we use the expression here, is giving an advantage to a member of a group that in the past was denied equal treatment, in particular, women and minorities. It "reverses" the historical order of preferences. The Supreme Court has largely forbidden the use of explicit numerical quotas for minorities, but it has not forbidden taking into account minority status

---

[15] Catherine A. MacKinnon, *Sexual Harassment of Working Women* (New Haven, CT: Yale University Press, 1978), pp. 1, 57–82.
[16] Ibid., p. 162.

in all contexts. The *weak form* consists of hiring a woman or a member of a minority over an equally qualified white male. The *strong form,* by contrast, consists of giving preference to women or minorities over better-qualified white males. Can such preference ever be justified? There are compelling arguments on both sides of the issue.[17]

Arguments favoring preferential treatment take three main forms. First, there is an argument based on compensatory justice: Past violations of rights must be compensated. Ideally such compensation should be given to individuals who in the past were denied jobs. But the costs and practical difficulties of determining such discrimination on a case-by-case basis through job-interviewing suggests instead giving preference on the basis of membership in a group that has been disadvantaged in the past. Second, sexism and racism still permeate our society today, and to counterbalance their insidious impact, reverse preferential treatment is warranted in order to ensure equal opportunity for minorities and women. Third, those utilitarians who favor reverse preferential treatment point to its good consequences: integrating women and minorities into the economic and social mainstream (especially in male-dominated professions like engineering), providing role models for minorities that build self-esteem, and strengthening corporate diversity.

Arguments against reverse preferential treatment condemn it as "reverse discrimination." It is a straightforward violation of the rights to equal opportunity of white males and others who are now not given a fair chance to compete on the basis of their qualifications. Granted, past violations of rights may call for compensation, but only compensation to specific individuals who are wronged and only in ways that do not violate the rights of others who did not personally wrong minorities. It is also permissible to provide special funding for educational programs for economically disadvantaged children, but not to use jobs as a compensatory device. Moreover, those utilitarians who are opposed to reverse preferential treatment point to its negative effects: lowering economic productivity by using criteria other than qualifications in hiring, encouraging racism by generating intense resentment generated among white males and their families, encouraging traditional stereotypes that minorities and women cannot make it on their own without special help, and thereby adding to self-doubts of members of these groups.

Various attempts have been made to develop intermediate positions sensitive to all the above arguments for and against

---

[17] Steven M. Cahn, ed., *The Affirmative Action Debate* (New York: Routledge, 1995); George E. Curry, ed., *The Affirmative Action Debate* (Reading, MA: Addison-Wesley, 1996).

strong preferential treatment. For example, one approach rejects blanket preferential treatment of special groups as inherently unjust, but permits reverse preferential treatment within companies that can be shown to have a history of bias against minorities or women. Another approach is to permit weak reverse preferential treatment but to forbid strong forms.

Discussion Topics

1. Present and defend your view as to whether preferential treatment of women and minorities is ever justified.

2. The majority of employers have adopted mandatory random drug testing on their employees, arguing that the enormous damage caused by the pervasive use of drugs in our society carries over into the workplace. Typically, the tests involve taking urine or blood samples, obtained under close observation, thereby raising questions about personal privacy as well as privacy issues about drug usage away from the workplace that is revealed by the tests. Present and defend your view concerning mandatory drug tests at the workplace.

   In your answer, take account of the argument set forth by Joseph R. DesJardins and Ronald Duska that, except where safety is a clear and present danger (as in the work of pilots, police, and the military), such tests are unjustified.[18] They contend that employers have a right to the level of performance for which they pay employees, a level typically specified in contracts and job descriptions. When a particular employee fails to meet that level of performance, then employers will take appropriate disciplinary action based on observable behavior. Either way, it is employee performance that is relevant in evaluating employees, not drug usage per se.

3. A company advertises for an engineer to fill a management position. Among the employees the new manager is to supervise is a woman engineer, Ms. X, who was told by her former boss that she would soon be assigned tasks with increased responsibility. The prime candidate for the manager's position is Mr. Y, a recent immigrant from a country known for confining the roles for women. Ms. X was alerted by other women engineers to expect unchallenging, trivial assignments from a supervisor with Mr. Y's background. Is there anything she can and should do? Would it be ethical for her to try to forestall the appointment of Mr. Y?

---

[18] Joseph DesJardins and Ronald Duska, "Drug Testing in Employment," *Business & Professional Ethics Journal* 6 (1987), pp. 3–21.

4. Jim Serra, vice president of engineering, must decide who to recommend for a new director-level position that was formed by merging the product (regulatory) compliance group with the environmental testing group.[19] The top inside candidate is Diane Bryant, senior engineering group manager in charge of the environmental testing group. Bryant is 36, exceptionally intelligent and highly motivated, and a well-respected leader. She is also five months pregnant and is expected to take an eight-week maternity leave two months before the first customer ship deadline (six months away) for a new product. Bryant applies for the job and, in a discussion with Serra, assures him that she will be available at all crucial stages of the project. Your colleague, David Moss, who is vice president of product engineering, strongly urges you to find an outside person, insisting that there is no guarantee that Bryant will be available when needed. Much is at stake. A schedule delay could cost several millions of dollars in revenues lost to competitors. At the same time, offending Bryant could lead her and perhaps other valuable engineers whom she supervises to leave the company. What procedure would you recommend in reaching a solution?

5. Engineering societies have generally portrayed participation by engineers in unions and collective bargaining in engineering as unprofessional and disloyal to employers. Thus, the NSPE code of ethics states, "Engineers shall not actively participate in strikes, picket lines, or other collective coercive action" (Sec. III, 1e). Critics reply that such generalized prohibitions reflect the excessive degree to which engineering is still dominated by corporations' interests. Discuss this issue with regard to the following case. Would the approach described below provide an effective alternative way of drawing attention to safety problems? What other options might be pursued, and would they still involve "collective coercive action"?

Management at a mining and refinery operation have consistently kept wages below industrywide levels. They have also sacrificed worker safety in order to save costs by not installing special structural reinforcements in the mines, and they have made no effort to control excessive pollution of the work environment. As a result, the operation has reaped larger-than-average profits. Management has been approached both by individuals and by representatives of employee groups about raising wages

---

[19] This case is a summary of Cindee Mock and Andrea Bruno, "The Expectant Executive and the Endangered Promotion," *Harvard Business Review*, January–February 1994, pp. 16–18.

and taking the steps necessary to ensure worker safety, but to no avail. A nonviolent strike is called and the metallurgical engineers support it for reasons of worker safety and public health.

> We need engineers with the courage to speak out when things are not right, and colleagues to support them when the need arises.
>
> —The authors

Whistleblowing and Loyalty

No topic in engineering ethics is more controversial than whistleblowing. A host of issues are involved. When is whistleblowing morally permissible? Is it ever morally obligatory, or is it beyond the call of duty? To what extent, if any, do engineers have a right to whistleblow, and when is doing so immoral and imprudent? When is whistleblowing an act of disobedience and disloyalty to an organization? What procedures ought to be followed in blowing the whistle? Before considering these questions, we briefly define *whistleblowing*. Then, after presenting two cases, we recommend procedures for responsible whistleblowing.

Definition

Whistleblowing occurs when an employee or former employee conveys information about a significant moral problem to someone in a position to take action on the problem, and does so outside regular in-house channels for addressing disputes or grievances. The definition has four main parts.

1. *Disclosure.* Information is intentionally conveyed outside approved organizational (workplace) channels or in situations where the person conveying it is under pressure from supervisors or others not to do so.
2. *Topic.* The information concerns what the person believes is a significant moral problem for the organization (or an organization with which the company does business). Examples of significant problems are serious threats to public or employee safety and well-being, criminal behavior, unethical policies or practices, and injustices to workers within the organization.
3. *Agent.* The person disclosing the information is an employee or former employee (or someone else closely associated with the organization).
4. *Recipient.* The information is conveyed to a person or organization in a position to act on the problem (as opposed, for example,

to telling it to a relative or friend who is in no position to do anything).[20] The desired response or "action" may consist in remedying the problem or merely alerting affected parties. Typically, though not always, the information being revealed is new or not fully known to the person or group receiving it.

Using this definition, we will speak of *external whistleblowing,* when the information is passed outside the organization. *Internal whistleblowing* occurs when the information is conveyed to someone within the organization (but outside approved channels or against pressures by immediate superiors to remain silent).

The definition also allows us to distinguish between open and anonymous whistleblowing. In *open whistleblowing,* individuals openly reveal their identity as they convey the information. *Anonymous whistleblowing,* by contrast, involves concealing one's identity. There are also overlapping cases that are partly open and partly anonymous, such as when individuals acknowledge their identities to a journalist but insist their names be withheld from anyone else.

Notice that the above definition does not mention the motives involved in the whistleblowing, and hence avoids assumptions about whether those motives are good or bad. Nor does it assume that the whistleblower is correct in believing there is a serious moral problem. In general, it leaves open the question of whether whistleblowing is justified. In turning to issues about justification, let us begin with two case studies, one in which the whistle was blown and one in which it was not.

Two Cases

Ernest Fitzgerald and the C-5A

One of the most publicized instances of open, external whistleblowing occurred on November 13, 1968. On that day, Ernest Fitzgerald was one of several witnesses called to testify before Senator William Proxmire's Subcommittee on Economy in Government concerning the C-5A, a giant cargo plane being built by Lockheed Aircraft Corporation for the Air Force. Fitzgerald, who had previously been an industrial engineer and management consultant, was then a deputy for management systems under the assistant secretary of the Air Force. During the preceding two years, he had reported huge cost overruns in the C-5A project to his superiors, overruns that by 1968 had hit $2 billion. He had argued forcefully against similar overruns in other projects, so forcefully that he had become unpopular with his superiors.

---

[20] We adopt the fourth condition from Marcia P. Miceli and Janet P. Near, *Blowing the Whistle: The Organizational and Legal Implications for Companies and Employees* (New York: Lexington Books, 1992), p. 15.

They pressured him not to discuss the extent of the C-5A over-runs before Senator Proxmire's committee. Yet when Fitzgerald was directly asked to confirm Proxmire's own estimates of the overruns on that November 13, he told the truth.

Doing so turned his career into a costly nightmare for himself, his wife, and his three children.[21] He was immediately stripped of his duties and assigned trivial projects, such as examining cost overruns on a bowling alley in Thailand. He was shunned by his colleagues. Within 12 days, he was notified that his promised civil service tenure was a computer error. And within four months, the bureaucracy was restructured so as to abolish his job. It took four years of extensive court battles before federal courts ruled that he had been wrongfully fired and ordered the Air Force to rehire him. After years of further litigation, involving fees of around $900,000, he was finally reinstated in his former position in 1981.

Dan Applegate and the DC-10

In 1974, the first crash of a fully loaded DC-10 jumbo jet occurred over the suburbs of Paris; 346 people were killed, a record for a single-plane crash. It was known in advance that such a crash was bound to occur because of the jet's defective design.[22]

The fuselage of the plane was developed by Convair, a sub-contractor for McDonnell Douglas. Two years earlier, Convair's senior engineer directing the project, Dan Applegate, had written a memo to the vice president of the company itemizing the dangers that could result from the design. He accurately detailed several ways the cargo doors could burst open during flight, depressurize the cargo space, and thereby collapse the floor of the passenger cabin above. Since control lines ran along the cabin floor, this would mean a loss of full control over the plane. Applegate recommended redesigning the doors and strengthening the cabin floor. Without such changes, he stated, it was inevitable that some DC-10 cargo doors would open in midair, resulting in crashes (Golich[22]).

In responding to this memo, top management at Convair disputed neither the technical facts cited by Applegate nor his predictions. Company officials maintained, however, that the possible

---

[21] Ernest Fitzgerald, *The High Priests of Waste* (New York: W. W. Norton, 1972); Berkeley Rice, *The C5-A Scandal* (Boston: Houghton-Mifflin, 1971).

[22] See John H. Fielder and Douglas Birsch, eds., *The DC-10 Case* (Albany, NY: State University of New York Press, 1992); Paul Eddy, Elaine Potter, and Bruce Page, *Destination Disaster* (New York: Quadrangle, 1976); John Godson, *The Rise and Fall of the DC-10* (New York: David McKay, 1975); John Newhouse, *The Sporty Game* (New York: Alfred A. Knopf, 1982). Vicki Golich: *The Political Economy of International Air Safety* (New York: St. Martin's Press, 1989), p. 75, 115.

financial liabilities Convair might incur prohibited them from passing on this information to McDonnell Douglas. These liabilities could be severe since the cost of redesign and the delay to make the necessary safety improvements would be very high and would occur at a time when McDonnell Douglas would be placed at a competitive disadvantage.

Moral Guidelines

Under what conditions can or should engineers blow the whistle? Certainly it should be done when substantial harm can result from an organization's acts of omission or commission. The harm may have occurred already but is not noticeable yet. It could also occur in the future. The harm may be done to a customer, to the general public, to workers, or to the shareholders. It may be in the form of faulty products, unsafe working conditions, unfair policies, or fraud. Nevertheless, because going outside one's organization with sensitive information is a serious undertaking, it stands to reason that certain conditions should be met before anyone blows the whistle:[23]

1. The actual or potential harm reported is serious and has been adequately documented;
2. The concerns have been reported to immediate superiors;
3. After not getting satisfaction from immediate superiors, regular channels within the organization have been used to reach up to the highest levels of management.

The information may then be released confidentially to a relevant government authority, and only when that fails to bring an adequate response should public disclosure be considered.

One needs to consider exceptions.[24] Condition (1) might not be met if it is difficult to obtain documentation because cloaks of secrecy are imposed on evidence that, if revealed, could supposedly aid commercial competitors or a nation's adversaries. In such cases it may be very difficult to establish adequate documentation and the whistleblowing would consist essentially of a request to the proper authorities to carry out an external investigation, or to request a court to issue an order for the release of information.

---

[23] Adapted from Richard T. De George, "Ethical Responsibilities of Engineers in Large Organizations: The Pinto Case," *Business and Professional Ethics Journal* 1 (Fall 1981), p. 6. De George also distinguishes between obligatory and mandatory whistleblowing.

[24] Gene G. James, "Whistle Blowing: Its Moral Justification," in *Business Ethics*, ed. W. Michael Hoffman and Jennifer Mills Moore (New York: McGraw-Hill, 1990), pp. 332–44.

Respect for Authority

Respect for authority is important in meeting organizational goals. Decisions must be made in situations where allowing everyone to exercise unrestrained individual discretion would create chaos. Moreover, clear lines of authority provide a means for identifying areas of personal responsibility and accountability.

The relevant kind of authority has been called *executive authority*: the corporate or institutional right given to a person to exercise power based on the resources of an organization.[30] It is distinguishable from *power* (or influence) in getting the job done. It is distinguishable, too, from *expert authority*: the possession of special knowledge, skill, or competence to perform some task or to give sound advice. Employees *respect authority* when they accept the guidance and obey the directives issued by the employer having to do with the areas of activity covered by the employer's institutional authority, assuming the directives are legal and do not violate norms of moral decency.

Taken together, loyalty, collegiality, and respect for authority create a presumption against whistleblowing, but it is a presumption that can be overridden. Loyalty, collegiality, and respect for authority are not excuses or justification for shielding irresponsible conduct. To think otherwise would be to lapse into a form of *corporate egoism*: the view that the corporation is more important than the wider good of the public. In addition to corporate virtues, there are public-oriented virtues, especially respect for the public's safety.

Protecting Whistleblowers

Most whistleblowers have suffered unhappy and often tragic fates. In the words of one lawyer who defended a number of them:

> Whistleblowing is lonely, unrewarded, and fraught with peril. It entails a substantial risk of retaliation which is difficult and expensive to challenge. Furthermore, "success" may mean no more than retirement to a job where the bridges are already burned, or monetary compensation that cannot undo damage to a reputation, career and personal relationships.[31]

Yet the vital service to the public provided by many whistleblowers has led increasingly to public awareness of a need to

---

[30] Joseph A. Pichler, "Power, Influence and Authority," in *Contemporary Management,* ed. Joseph W. McGuire (Englewood Cliffs, NJ: Prentice Hall, 1974), p. 428; Richard T. De George, *The Nature and Limits of Authority* (Lawrence: University Press of Kansas, 1985).

[31] Peter Raven-Hansen, "Dos and Don'ts for Whistle-Blowers: Planning for Trouble," *Technology Review* 82 (May 1980), p. 44.

*"One final question. As far as you know, have you any family
history of loose-cannonism or whistle-blowing?"*

protect them against retaliation by employers. Government
employees have won important protections. Various federal laws
related to environmental protection and safety and the Civil Service Reform Act of 1978 protect them against reprisals for lawful
disclosures of information believed to show "a violation of any
law, rule, or regulation, mismanagement, a gross waste of funds,
an abuse of authority, or a substantial and specific danger to
public health and safety."[32] The fact that few disclosures are
made appears to be due mostly to a sense of futility—the feeling
that no corrective action will be undertaken or that many years
may lapse before a case is closed satisfactorily. In the private sec-
tor, employees are covered by statutes forbidding firing or
harassing of whistle-blowers who report to government regulatory agencies the violations of some 20 federal laws, including
those covering coal mine safety, control of water and air pollution, disposal of toxic substances, and occupational safety and

---

[32] Ibid., p. 42; Stephen H. Unger, *Controlling Technology: Ethics and the
Responsible Engineer,* 2nd ed. (New York: Holt, Rinehart and Winston, 1992),
pp. 179–81.

health. In a few instances, unions provide further protection. Overall, the laws concerning whistleblowing are in transition, and a number of observers believe they are moving in directions favorable to responsible whistleblowing.[33]

Nevertheless, there is still one group of employees who are caught in particularly difficult positions because they are paid by private companies to work on government projects at government sites. Nowhere has remedial action been resisted and delayed as much as on clean-up jobs at nuclear sites across the U.S. It is not unusual that the management of the contractor or of the on-site supervising government agency, or both, will shroud complaints in an official cloak of secrecy "for national security reasons" while doing little to resolve the underlying problems. The Government Accountability Project (GAP, website www.whistleblower.org) is an independent organization that offers its assistance to contractor employees and government employees caught in such situations.

In addition to laws that protect government employees, there is legislation to reward whistleblowers who report overcharging on federal contracts (see Discussion Topic 3).

Such laws, when they are carefully formulated and enforced, provide two types of benefits for the public, in addition to protecting the responsible whistleblower: episodic and systemic. The *episodic* benefits are in helping to prevent harm to the public in particular situations. The *systemic* benefits are in sending a strong message to industry to act responsibly or be subject to public scrutiny once the whistle is blown. While the law provides a measure of protection to the responsible whistleblower, there is also an important potential role for professional societies. Until recently, few societies openly supported engineers who had followed their codes of ethics in notifying "proper authorities" after their superiors had overruled them and their professional judgments about dangers to the public. But this is changing. For example, the Institute of Electrical and Electronics Engineers (IEEE) has supported responsible whistleblowers by backing them in court and by establishing forms of honorary recognition for whistleblowers who act according to its ethical code, and by helping to locate new jobs for discharged engineers.[34] Another

---

[33] Kenneth Walters, "Your Employees' Right to Blow the Whistle," *Harvard Business Review* 53 (July 1975), p. 34; David W. Ewing, *Freedom Inside the Organization* (New York: McGraw-Hill, 1977), p. 113; Alan F. Westin, ed., *Whistle-Blowing! Loyalty and Dissent in the Corporation* (New York: McGraw-Hill, 1981), pp. 163–64; James C. Petersen and Dan Farrell, *Whistle-Blowing* (Dubuque, IA: Kendall/Hunt, 1986), p. 20.

[34] Robert M. Anderson, Robert Perrucci, Dan E. Schendel, and Leon E. Trachtman, *Divided Loyalties* (West Lafayette, IN: Purdue University Press, 1980).

avenue of protection for engineers being explored by professional societies is the publication in their journals of the names of companies who take unjust reprisals against whistleblowers. Some societies have experimented with hot-lines for engineers contemplating blowing the whistle, but caution advised by lawyers who thought threats of lawsuits might require prohibitively expensive insurance coverage caused at least one society, the IEEE, to abandon its promising hot-line prematurely.

Commonsense Procedures

It is clear that a decision to whistleblow, whether within or outside an organization, is a serious matter that deserves careful reflection. And there are several rules of practical advice and common sense that should be heeded before taking this action.[35]

1. Except for extremely rare emergencies, always try working first through normal organizational channels. Get to know both the formal and informal (unwritten) rules for making appeals within the organization.

2. Be prompt in expressing objections. Waiting too long may create the appearance of plotting for your advantage and seeking to embarrass a supervisor.

3. Proceed in a tactful, low-key manner. Be considerate of the feelings of others involved. Always keep focused on the issues themselves, avoiding any personal criticisms that might create antagonism and deflect attention from solving those issues.

4. As much as possible, keep supervisors informed of your actions, both through informal discussion and formal memorandums.

5. Be accurate in your observations and claims, and keep formal records documenting relevant events.

6. Consult trusted colleagues for advice—avoid isolation.

7. Before going outside the organization, consult the ethics committee of your professional society. For employees of the U.S. Government and its contractors, the Government Accountability Project may be a good source (www.whistleblower.org).

8. Consult a lawyer concerning potential legal liabilities.

---

[35] Stephen H. Unger, "How to be Ethical and Survive," *IEEE Spectrum* 16 (December 1979), pp. 56–57; Frederick Elliston, John Keenan, Paula Lockhart, and Jane van Schaick, *Whistle-Blowing Research: Methodological and Moral Issues* (New York: Praeger, 1985); Frederick Elliston, John Keenan, Paula Lockhart, and Jane van Schaick, *Whistle-Blowing: Managing Dissent at the Workplace* (New York: Praeger, 1985).

Beyond Whistleblowing

Sometimes whistleblowing is a practical moral necessity. But generally it holds little promise as the best possible method for remedying problems and should be viewed as a last resort.

The obvious way to remove the need for internal whistleblowing is for management to allow greater freedom and openness of communication within the organization. By making those channels more flexible and convenient, the need to violate them would be removed. But this means more than merely announcing formal "open-door" policies and appeals procedures that give direct access to higher levels of management. Those would be good first steps, and a further step would be the creation of an ombudsperson or an ethics review committee with genuine freedom to investigate complaints and make independent recommendations to top management. The crucial factor that must be involved in any structural change, however, is the creation of an atmosphere of positive affirmation of engineers' efforts to assert and defend their professional judgments in matters involving ethical considerations.

What about external whistleblowing? Much of it can also be avoided by the same sorts of intra-organizational modifications. Yet there will always remain troublesome cases where top management and engineers differ in their assessments of a situation even though both sides may be equally concerned to meet their professional obligations to safety. To date, the assumption has been that management has the final say in any such dispute. But our view is that engineers have a right to some further recourse in seeking to have their views heard, including confidential discussions with the ethics committees of their professional societies.

When an engineer so strongly disagrees with the purposes of a product, the policies of management, the low level of safety in manufacturing/construction, or the lack of candor in advertising, he or she may simply decide to quit. Under such circumstances engineers may ask to be removed from the projects at hand, or they may decide to separate entirely from their employer. Freed of the usual employment obligations, the now unemployed engineer can more freely blow the whistle but should keep in mind that this may lessen chances of finding employment in the future, especially when claims of wrongdoing are greatly exaggerated, whether by the whistleblower or a news medium. One obligation carried over from the resigned position is the duty not to divulge trade or national security secrets. When that is unavoidable, the secrets should be revealed only in a manner the law may allow or the engineer's professional society recommends.

The following examples are but a few of the many that could be cited and do not include engineers or scientists who abstained from certain lines of work because they could tell from the outset that their principles might be compromised. Norbert Wiener, the father of cybernetics, is one such person. He refused to work on projects which could not be freely discussed and which would threaten human well-being and liberties.

Roy Woodruff was associate director of defense systems at the Lawrence Livermore National Laboratory when he resigned his position over differences with Edward Teller, who was his boss and Director of the Laboratory. Woodruff declared that tests had proven an X-ray laser beam, the Excalibur system proposed for use in the Star Wars Project, to be ineffective as a tool for disabling incoming enemy missiles. Teller, on the other hand, extolled its virtues and had the ear of President Reagan and top level decision makers in Washington. Woodruff resigned his position as associate director but stayed on at the Laboratory to pursue other interests while trying to get more of a hearing for his assessment of space-based weaponry.[36]

David Parnas, a computer scientist, was also involved in a Star Wars project. He resigned from an advisory panel on computing when he lost his initial enthusiasm for the Strategic Defense Initiative (SDI) after only the first meeting of the panel. When agency officials would not seriously listen to his doubts about the feasibility of the project, he gradually succeeded through journal articles, open debates, and public lectures to convince the profession that Star Wars did not differ much from conventional anti-ballistic-missile defense without overcoming earlier shortcomings. Indeed, the system's complexity made it practically impossible to write software as reliable as it ought to be in tight-trigger situations. For his efforts on behalf of the public interest, he was honored with the Norbert Wiener Award by the society of Computer Professionals for Social Responsibility (CPSR).[37] Incidentally, Parnas also chafed at what he saw as opportunistic proposals by researchers in industry and academe for favorite projects often only remotely related to SDI, but with large budgets supported by a generous pool of available money, thus providing SDI with some base of support.

Greg Minor, Richard Hubbard, and Dale Bridenbaugh were nuclear reactor specialists with General Electric. Independently of each other they had found a variety of safety defects in GE reactors but received no responses from management that would have allayed their concerns. After the three engineers had individually decided to resign, they quit in unison so their step would

---

[36] Robert Scheer, "The Man Who Blew the Whistle on 'Star Wars'," *Los Angeles Times Magazine,* July 17, 1988, p. 6–32.
[37] Carl Page, "Star Wars, Down but Not Out," Fall 1996 Newsletter of Computer Professionals for Social Responsibility, v.14, n.4, Fall 1996.

draw greater attention. Thereafter they proceeded to advise citizens' groups and the Union of Concerned Scientists (UCS) on nuclear plant safety. In 1978 they gave technical advice for the filming of *The China Syndrome* shortly before the Three Mile Island accident occurred (unrelated to GE reactors).[38]

Discussion Topics

1. According to Kenneth Kipnis, a professor of philosophy, Dan Applegate and his colleagues share the blame for the death of the passengers in the DC-10 crash. Kipnis contends that the engineers' overriding obligation was to obey the following principle: "Engineers shall not participate in projects that degrade ambient levels of public safety unless information concerning those degradations is made generally available."[39] Do you agree or disagree with Kipnis, and why? Was Applegate obligated to blow the whistle?

2. Present and defend your view as to whether in the case described below the actions of Ms. Edgerton and her supervisor were morally permissible, obligatory, or admirable. Did Ms. Edgerton have a professional moral right to act as she did? Was hers a case of legitimate whistle-blowing?

   In 1977, Virginia Edgerton was senior information scientist on a project for New York City's Criminal Justice Coordinating Council. The project was to develop a computer system for use by New York district attorneys in keeping track of data about court cases. It was to be added on to another computer system, already in operation, that dispatched police cars in response to emergency calls. Ms. Edgerton, who had 13 years of data-processing experience, judged that adding on the new system might result in overloading the existing system in such a way that the response time for dispatching emergency vehicles might be increased. Because it might risk lives to test the system in operation, she recommended that a study be conducted ahead of time to estimate the likelihood of such overload.

   She made this recommendation to her immediate supervisor, the project director, who refused to follow it. She then sought advice from the IEEE, of which she was a member. The Institute's Working Group on Ethics and Employment Practices referred her to the manager of systems programming at Columbia University's computer center, who verified that she was raising a legitimate issue.

---

[38] Karen Fitzgerald, "Whistleblowing: Not Always a Losing Game," IEEE Spectrum, December 1990, p. 49–52.

[39] Kenneth Kipnis, "Engineers Who Kill: Professional Ethics and the Paramountcy of Public Safety," *Business and Professional Ethics Journal* 1 (1981), p. 82.

Next she wrote a formal memo to her supervisor, again requesting the study. When her request was rejected, she sent a revised version of the memo to New York's Criminal Justice Steering Committee, a part of the organization for which she worked. In doing so, she violated the project director's orders that all communications to the Steering Committee be approved by him in advance. The project director promptly fired her for insubordination. Later he stated: "It is . . . imperative that an employee who is in a highly professional capacity, and has the exposure that accompanies a position dealing with top level policy makers, follow expressly given orders and adhere to established policy."[40]

3. A controversial area of recent legislation allows whistleblowers to collect money. Federal tax legislation, for example, pays informers a percentage of the money recovered from tax violators. And the 1986 False Claims Amendment Act allows 15 to 25 percent of the recovered money to go to whistle-blowers who report overcharging in federal government contracts to corporations. These sums can be substantial because lawsuits can involve double and triple damages as well as fines. Discuss the possible benefits and drawbacks of using this approach in engineering and specifically concerning safety matters. Is the added incentive to whistle-blow worth the risk of encouraging self-interested motives in whistleblowing?

4. Do you see any special moral issues raised by anonymous whistleblowing?[41]

5. An engineer likes the job at the manufacturing plant, likes the colleagues and lower level managers, but is concerned about the dangerous defects which have only now been discovered in a product after an entire production run has already reached the distributors. Upper level management at headquarters is not taking any steps to recall the product or halt its distribution. If the engineer were now to inform public authorities, this could hurt the plant's reputation and balance sheet, it could lead to a reduction in the workforce, and if the engineer's role were to become public (or he had blown the whistle openly), the engi-

---

[40] "Edgerton Case," Reports of the IEEE-CSIT Working Group on Ethics & Employment and the IEEE Member Conduct Committee in the matter of Virginia Edgerton's dismissal as information scientist of New York City. Reproduced in *Technology and Society* 22 (June 1978), pp. 3–10. See also *The Institute,* news supplement to *IEEE Spectrum,* June 1979, p. 6, for articles on her IEEE Award for Outstanding Public Service.

[41] Frederick A. Elliston, "Anonymous Whistle-Blowing," *Business and Professional Ethics Journal* 1 (1982), pp. 39–58.

personal

neer's family would lose the income on which it depends. Is the engineer facing (a) a conflict of interests, (b) conflicting interests, (c) divided loyalties, or (d) a combination of the former?

two ~~three~~ aspects involved

legal

6. Analyze the issues involved in the whistleblowing case described as Topic 5 above from the standpoints of the various ethical theories discussed in Chapter 2: utilitarianism, duty theories, rights theories, and virtue theories.

7. When an engineer blows the whistle in a responsible way as the text has described it, would this be an act based on (i) a right, (ii) a moral duty, or (iii) both? When you now think of the engineer not in a professional role, but as (a) a human being, or (b) a concerned citizen, would your answers differ from the ones you gave earlier?

8. In what ways might the relationship between a company and a consulting engineer be likened to that between a company and an employed engineer? When a company hires a consulting engineer (another company or an individual consultant), could the consultant be viewed by a company manager as just another employee who must be paid? On the other hand, shouldn't the consultant consider the company as his or her client? Where does that leave the clients whose project the consultant is working on as part of his services to the company? Describe what you think the loyalty priorities for each of the parties are or should be.

9. Today many engineers have lost permanent employment due to the current trend of "downsizing" in industry. Some of these engineers are hired again on a temporary basis as "contract engineers." A contract engineer may receive more in cash during a pay period than a fully employed engineer may get, but that is so because benefits such as holiday pay, sick leave, health insurance, and retirement contribution are usually excluded and the company retains the opportunity to dismiss the engineer at the end of the contract. Contracts can be made to last only briefly but may have renewal options. Reexamine the discussion topic above with a contract engineer assuming the role of consulting engineer. You may wish to explore both discussion topics further by talking to a consulting engineer, or a lawyer advising consulting and contract engineers, but first make sure you do not have to pay for this consultation. You may find them at a meeting of the professional society representing your chosen field, and you may even get a free or reduced-cost meal if you are a student, courtesy of the society.

6

Global Issues

An American family purchasing a General Motors Pontiac Le Mans in 1990 probably believed their purchase would help American auto workers far more so than would the purchase of a foreign-made car. According to Labor Secretary Robert Reich's estimates, however, only $4,000 of the $10,000 sticker price would go directly to Americans. Indeed, that $4,000 would not go to American assembly-line workers, but instead mostly to Detroit strategists in higher management, New York bankers and attorneys, insurance workers spread throughout the country, and General Motors shareholders who include both Americans and non-American foreign investors. The remaining $6,000 would be distributed as follows: "about $3,000 goes to South Korea for routine labor and assembly operations, $1,750 to Japan for advanced components (engines, transaxles, and electronics), $750 to West Germany for styling and design components, $250 to Britain for advertising and marketing services, and about $50 to Ireland and Barbados for data processing."[1] A president of GM once remarked that what is good for GM is good for America, but apparently, in this case, it is even better for the rest of the world.

As both workers and consumers, all of us live in an increasingly international marketplace. American and foreign consumers alike will force companies to compete within a global market. Politically, too, our lives are interwoven with those of people around the world. Our existence as a human species depends on sustaining an increasingly fragile environment. As we enter the twenty-first century, our economic well-being, national security, and biological existence are interdependent with other nations in ways that only a soothsayer could have foreseen a century ago.

---

[1] Robert B. Reich, *The Work of Nations* (New York: Vintage Books, 1992), p. 113.

The word *global* in this chapter's title refers to both the international context of engineering and the increasingly pervasive social and environmental dimensions of engineers' work. As responsible social experimenters, engineers need to take these dimensions into account in making engineering decisions and career choices. We will explore these dimensions by discussing three topics: multinational corporations, environmental ethics, and weapons development.

Multinational Corporations

Multinational corporations do extensive business in more than one country. In some cases, the operations of corporations are spread so thinly around the world that the location of their official headquarters in any one home country is largely incidental and essentially a matter of historical circumstance. The benefits to U.S. companies of doing business in less economically developed countries are clear: inexpensive labor, availability of natural resources, favorable tax arrangements, and fresh markets for products. The benefits to the participants in developing countries are equally clear: new jobs, jobs with higher pay and greater challenge, transfer of advanced technology, and an array of social benefits from sharing wealth.

Yet moral difficulties arise, along with business and social complications. Who loses jobs at home when manufacturing is taken "offshore"? What does the host country lose in resources, control over its own trade and standards, and political independence? To what extent are "multinationals" influencing and even usurping the roles of national governments as they play off one government against another? (For instance, see Discussion Topic 6.) And what are the moral responsibilities of corporations and individuals operating in less economically developed countries? Here we focus on the last question. Before doing so, we think it helpful to introduce the concepts of technology transfer and appropriate technology.

Technology Transfer and Appropriate Technology

Technology transfer is the process of moving technology to a novel setting and implementing it there.[2] Technology includes both hardware (machines and installations) and technique (technical, organizational, and managerial skills and procedures). A novel setting is any situation containing at least one new variable relevant to the success or failure of a given technology. The setting may be within a country where the technology is already

---

[2] Peter B. Heller, *Technology Transfer and Human Values* (New York: University Press of America, 1985), p. 119.

used elsewhere, or a foreign country, which is our present interest. A variety of agents may conduct the transfer of technology: governments, universities, private volunteer organizations, consulting firms, and multinational corporations.

In most instances, the transfer of technology from a familiar to a new environment is a complex process. The technology being transferred may be one that originally evolved over a period of time and is now being introduced as a ready-made, completely new entity into a different setting. Discerning how the new setting differs from familiar contexts requires the imaginative and cautious vision of "cross-cultural social experimenters."

The expression *appropriate technology* is widely used, but with a variety of meanings. We use it in a generic sense to refer to identification, transfer, and implementation of the most suitable technology for a new set of conditions. Typically the conditions include social factors that go beyond routine economic and technical engineering constraints. Identifying them requires attention to an array of human values and needs that may influence how a technology affects the novel situation. Thus, "appropriateness may be scrutinized in terms of scale, technical and managerial skills, materials/energy (assured availability of supply at reasonable cost), physical environment (temperature, humidity, atmosphere, salinity, water availability, etc.), capital opportunity costs (to be commensurate with benefits), but especially human values (acceptability of the end-product by the intended users in light of their institutions, traditions, beliefs, taboos, and what they consider the good life)."[3]

Examples include the introduction of agricultural machines and long-distance telephones. A country with many poor farmers can make better immediate use of small, single- or two-wheel tractors that can serve as motorized plows, to pull wagons or to drive pumps, than it can of huge diesel tractors that require collectivized or agribusiness-style farming. On the other hand, the same country may benefit more from the latest in wireless communication technology to spread its telephone service to more people and over long distances than it can from old-fashioned transmission by wire.

Appropriate technology also implies that the technology should contribute to and not distract from *sustainable* development of the host country by providing for careful stewardship of its natural resources and not degrading the environment beyond its carrying capacity.

Appropriate technology overlaps with, but is not reducible to, *intermediate technology,* which lies between the most advanced

---

[3] Ibid.

forms available in industrialized countries and comparatively primitive forms in less-developed countries.[4] The British economist E. F. Schumacher argued that intermediate technologies are preferable because the most advanced technologies usually have harmful side effects, such as causing mass migrations from rural areas to cities where corporations tend to locate. These migrations cause overcrowding, and with it poverty, crime, and disease. Far more appropriate, he argued, are smaller-scale technologies replicated throughout a less-developed country, using low capital investment, labor intensiveness to provide needed jobs, local resources where possible, and simpler techniques manageable by the local population given its education facilities.

We mention intermediate technology, and the movement inspired by Schumacher, not to offer a general endorsement (often it has been dramatically beneficial; at other times, not particularly effective), but to emphasize that it is only one conception of appropriate technology. "Appropriate technology" is a generic concept that applies to all attempts to emphasize wider social factors when transferring technologies. As such, it reinforces and amplifies our view of engineering as social experimentation.

With these distinctions in mind, let us turn in some detail to a case study illustrating the complexities of engineering within multinational settings.

Bhopal

Union Carbide in 1984 operated in 37 "host countries" in addition to its "home country," the United States, ranking 35th in size among U.S. corporations. On December 3, 1984, the operators of Union Carbide's plant in Bhopal, India, became alarmed by a leak and overheating in a storage tank. The tank contained methyl isocyanate (MIC), a toxic ingredient used in pesticides. As a concentrated gas, MIC burns any moist part of bodies with which it comes in contact, scalding throats and nasal passages, blinding eyes, and destroying lungs. Within an hour, the leak exploded in a gush that sent 40 tons of deadly gas into the atmosphere.[5] The result was the worst industrial accident in history: 500,000 persons exposed to the gas, 2500 deaths within a few days, 10,000 permanently disabled, 100,000 others injured, and a great loss of livestock. By 10 years later, 12,000 death claims and 870,000 personal injury claims had been submitted, but only $90 million of Union Carbide's $470 million settlement amount had been distributed.

---

[4] E. F. Schumacher, *Small Is Beautiful* (New York: Harper & Row, 1973).

[5] Paul Shrivastava, *Bhopal, Anatomy of a Crisis* (Cambridge, MA: Ballinger, 1987). See also Gary Stix, "Bhopal: A Tragedy in Waiting," *IEEE Spectrum*, June 1989, pp. 47–50; Larry Everest, *Behind the Poison Cloud: Union Carbide's Bhopal Massacre* (Chicago: Banner, 1985).

The disaster was caused by a combination of extremely lax safety procedures, gross judgment errors by local plant operators, and possible sabotage with unintended consequences. In retrospect, it is clear that greater sensitivity to social factors was needed in transferring chemical technology to a country foreign to the supplier of the technology. The extent of the disaster would have been lessened, however, if Union Carbide had designed the plant with smaller (though more) tanks to store MIC as it had been required to do in France.

The government of India required the Bhopal plant to be operated entirely by Indian workers. Hence, Union Carbide at first took admirable care in training plant personnel, flying them to its West Virginia plant for intensive training. It also had teams of U.S. engineers make regular on-site safety inspections. But in 1982, financial pressures led Union Carbide to relinquish its supervision of safety at the plant, but it retained general financial and technical control. The last inspection by a team of U.S. engineers occurred that year, two years before the explosion, despite the fact that the team had warned of many of the hazards that contributed to the disaster.

During the years after 1982, safety practices eroded. One source of the erosion was personnel: high turnover of employees, failure to properly train new employees, and low technical preparedness of the local labor pool. Workers handling pesticides, for example, learned more from personal experience than from study of safety manuals about the dangers of the pesticides. But even after suffering chest pains, vomiting, and other symptoms, they would sometimes fail to wear safety gloves and masks because of high temperatures caused by lack of air-conditioning in the plant.

The other source of eroding safety practices was the move away from U.S. standards (contrary to Carbide's written policies) toward lower Indian standards. By December of 1984, several extreme hazards, in addition to many smaller ones, were present. (See Figure 6–1.)

First, the tanks storing the MIC gas were overloaded. Carbide's manuals specified they were never to be filled to more than 60 percent of capacity; in emergencies, the extra space could be used to dilute the gas. The tank that caused the problem was in fact more than 75 percent full.

Second, a standby tank that was supposed to be kept empty for use as an emergency dump tank already contained a large amount of the chemical.

Third, the tanks were supposed to be refrigerated to make the chemical less reactive if trouble should arise. But the refrigeration unit had been shut down five months before the accident as a cost-cutting measure, making tank temperatures three to four times what they should have been.

According to one account, a disgruntled employee unscrewed a pressure gauge on a storage tank and inserted a hose into it. He knew and intended that the water he poured into the tank would do damage, but he did not know it would cause such immense damage. According to another account, a relatively new worker had been instructed by a new supervisor to flush out some pipes and filters connected to the chemical storage tanks. Apparently the worker properly closed valves to isolate the tanks from the pipes and filters being washed, but he failed to insert the required safety disks to back up the valves in case they leaked. (He knew that valves leaked, but he did not check for leaks: "It was not my job." The safety disks were the responsibility of the maintenance department, and the position of second-shift supervisor had been eliminated.) Lawyers and their experts whose tasks include pinpointing blame, and engineers who design similar plants, need to know exactly what caused the pressure in the tank to rise, but

Figure 6–1
Diagram of Bhopal system. *(From Ward Worthy, "Methyl Isocyanate: The Chemistry of a Hazard," C & EN, February 11, 1985.)*

Reproduced with permission from *Chemical Engineering News,* Feb. 11, 1985, *63* (66), p. 29. Copyright 1985 American Chemical Society.

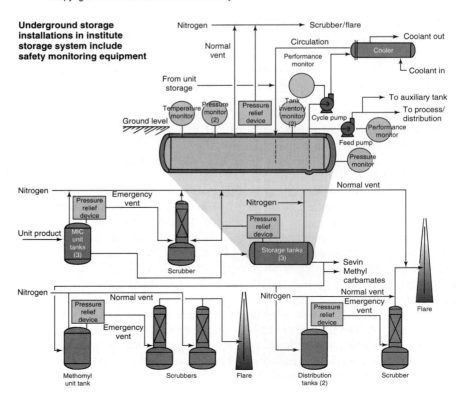

for the purposes of our present discussion it is sufficient to be concerned about the subsequent failure of the plant's safety systems. By the time the workers noticed a gauge showing the mounting pressure and began to feel the sting of leaking gas, they found their main emergency procedures unavailable. The primary defense against gas leaks was a vent-gas scrubber designed to neutralize the gas. It was shut down (and was turned on too late to help), because it was assumed to be unnecessary during times when production was suspended.

The second line of defense was a flare tower that would burn off escaping gas missed by the scrubber. It was inoperable because a section of the pipe connecting it to the tank was being repaired. Finally, workers tried to minimize damage by spraying water 100 feet into the air. The gas, however, was escaping from a stack 120 feet high.

Within two hours, most of the chemicals in the tank had escaped to form a deadly cloud over hundreds of thousands of people in Bhopal. As was common in India, desperately poor migrant laborers had become squatters—by the tens of thousands—in the vacant areas surrounding the plant. They had come with hopes of finding any form of employment, as well as to take advantage of whatever water and electricity was available.

Virtually none of the squatters had been officially informed by Union Carbide or the Indian government of the danger posed by the chemicals being produced next door to them. (The only voice of caution was that of a concerned journalist, Rajukman Keswani, who had written articles on the dangers of the plant and had posted warnings: "Poison Gas. Thousands of Workers and Millions of Citizens are in Danger.") There had been no emergency drills, and there were no evacuation plans: The scope of the disaster was greatly increased because of total unpreparedness.

"When in Rome"

What, in general, are the moral responsibilities of multinational corporations, like Union Carbide and General Motors, and their engineers? One tempting view is that corporations and employees are merely obligated to obey the laws and dominant customs of the host country: "When in Rome do as the Romans do." This view is a version of *ethical relativism,* the claim that actions are morally right within a particular society when (and only because) they are approved by law, custom, or other conventions of that society. Ethical relativism, however, is false because it implies moral absurdities. For example, it would justify horrendously low safety standards, if that were all a country required. Laws and conventions are not morally self-certifying. Instead, they are always open to criticism in light of moral reasons concerning human rights, the public good, duties to respect people, and virtues.

An opposite view would have corporations and engineers retain precisely the same practices endorsed at home, never making any adjustments to a new culture. This view is a version of *ethical absolutism,* the claim that moral principles have no justified exceptions and that what is morally true in one situation is true everywhere else. Absolutism is false because it fails to take account of how moral principles can come into conflict, forcing some justified exceptions. Absolutism also fails to take account of the many variable facts.

These considerations led us in Chapter 2 to endorse *ethical relationalism:* Moral judgments are and should be made in relation to factors that vary from situation to situation, usually making it impossible to formulate rules that are both simple and absolute. Moral judgments are contextual in that they are made in relation to a wide variety of factors—including the customs of other cultures. Note that relationalism only says that foreign customs are morally relevant. It does not say they are automatically decisive or self-authoritative in determining what should be done. This crucial difference sets it apart from ethical relativism.

Relationalism, we should add, is also consistent with *ethical pluralism,* the view that there is more than one justifiable moral perspective. In particular, there may be a number of morally permissible variations in formulating, interpreting, and applying basic moral principles. Not all rational and morally concerned people must see all specific moral issues the same way. This is as true in thinking about multinational corporations as it is in more everyday issues where we recognize that reasonable people can see moral issues differently and still be reasonable.

International Rights

If moral values are open to alternative interpretations, are there nevertheless some minimal standards that must be met? Let us respond to this question within the framework of rights ethics. A human right, by definition, is a moral entitlement that places obligations on other people to treat one with dignity and respect. If it makes sense at all, it makes sense across cultures, thereby providing a standard of minimally decent conduct that corporations and engineers are morally required to meet.

How can this general doctrine of human rights be applied practically, to help us understand the responsibilities of corporations doing business in other countries? In a pioneering book, *The Ethics of International Business,* Thomas Donaldson formulates a list of "international rights," human rights that are implied by, but more specific than, the most abstract human rights to liberty and fairness. These international rights have great importance and are often put at risk. Their exact requirements must be understood contextually, depending on the particular traditions and economic resources available in particular societies. Just as "ought implies can," rights do not require the impossible, and they also apply only within structured societies that provide a framework for understanding how to fairly distribute the burdens associated with them.

Donaldson suggests there are 10 such international rights.[6]

1. The right to freedom of physical movement.
2. The right to ownership of property.
3. The right to freedom from torture.
4. The right to a fair trial.
5. The right to nondiscriminatory treatment (freedom from discrimination on the basis of such characteristics as race or sex).
6. The right to physical security.
7. The right to freedom of speech and association.
8. The right to minimal education.
9. The right to political participation.
10. The right to subsistence.

These are human rights; as such, they place restrictions on how multinational corporations may act in other societies, even when those societies do not recognize the rights in their laws and customs. For example, the right to nondiscriminatory treatment would make it wrong for corporations to participate in discrimination against women and racial minorities even though this may be a dominant custom in the host country. Again, the right to physical security requires supplying protective goggles to workers running metal lathes, even when this is not required by the laws of the host country.

Although these rights have many straightforward cross-cultural applications, they nevertheless need to be applied contextually to take into account some features of the economy, laws, and customs of host countries. Not surprisingly, many difficulties and gray areas arise. One type of problem concerns the level of stringency required in matters such as degrees of physical safety at the workplace. Workers in less economically developed countries are often willing to take greater risks than would be acceptable to workers in the United States. Here Donaldson recommends

---

[6] Thomas Donaldson, *The Ethics of International Business* (New York: Oxford University Press, 1989), p. 81.

applying a "rational empathy test" to determine if it is morally permissible for a corporation to participate in the practices of the host country: Would citizens of the home country find the practice acceptable if their home country were in circumstances economically similar to those of the host country? For example, in determining whether a certain degree of pollution is acceptable for a U.S. company with a manufacturing plant located in India, the U.S. company would have to decide whether the pollution level would be acceptable under circumstances where the United States had a comparable level of economic development.

A second, quite different, type of problem arises where the practice is not so directly linked to economic factors, as in racial discrimination. Here Donaldson insists that unless one can do business in the country without engaging in practices that violate human rights, then corporations must simply leave and go to other countries.

Discussion Topics

1. Following the disaster at Bhopal, Union Carbide argued that officials at its U.S. corporate headquarters had no knowledge of the violations of Carbide's official safety procedures and standards. This has been challenged as documents were uncovered showing they knew enough to have warranted inquiry on their part, but let us assume they were genuinely ignorant. Would ignorance free them of responsibility for all aspects of the disaster?

2. Export of hazardous technologies, such as the manufacture of asbestos, to less-developed countries is motivated in part by cheaper labor costs, but another factor is that workers are willing to take greater risks. How does Donaldson's view apply to this issue?

   Also, do you agree with Richard De George's view that taking advantage of this willingness need not be unjust exploitation if several conditions are met: (1) Workers are informed of the risks. (2) They are paid more for taking the risks. (3) The company takes some steps to lower the risks, even if not to the level acceptable for U.S. workers.[7]

   How would you assess Union Carbide's handling of worker safety? Take into account the remarks of an Indian worker interviewed *after* the disaster. The worker was then able to stand

---

[7] Richard T. De George, "Ethical Dilemmas for Multinational Enterprise: A Philosophical Overview," in *Ethics and the Multinational Enterprise,* ed. W. Michael Hoffman, Ann E. Lange, and David A. Fedo (New York: University Press of America, 1986).

only a few hours each day because of permanent damage to his lungs. During that time, he begged in the streets while he awaited his share of the legal compensation from Union Carbide. When asked what he would do if offered work again in the plant knowing what he knew now, he replied: "If it opened again tomorrow I'd be happy to take any job they offered me. I wouldn't hesitate for a minute. I want to work in a factory, any factory. Before 'the gas' [disaster] the Union Carbide plant was the best place in all Bhopal to work."[8]

3. During 1972 and 1973, the president of Lockheed, A. Carl Kotchian, authorized secret payments totalling around $12 million beyond a contract to representatives of Japan's Prime Minister Tanaka. Later revelations of the bribes helped lead to the resignation of Tanaka and also to new laws in this country forbidding such payments. In 1995, long after Tanaka's death, the agonizingly slow trial and appeals process came to an end as Japan's Supreme Court reaffirmed the guilty verdicts, but so far no one has been jailed and the case appears to have had little recent impact on business and politics in Japan.

   Mr. Kotchian believed at that time it was the only way to assure sales of Lockheed's TriStar airplanes in a much-needed market. In explaining his actions, Mr. Kotchian cited the following facts.[9] (1) There was no doubt in his mind that the only way to make the sales was to make the payments. (2) No U.S. law at the time forbade the payments. (3) The payments were financially worthwhile, for they totaled only 3 percent of an expected $430 million income for Lockheed. (4) The sales would prevent Lockheed layoffs, provide new jobs, and thereby benefit workers' families and their communities as well as the stockholders. (5) He himself did not personally initiate any of the payments, which were all requested by Japanese negotiators. (6) In order to give the TriStar a chance to prove itself in Japan, he felt he had to "follow the functioning system" of Japan. That is, he viewed the secret payments as the accepted practice in Japan's government circles for this type of sale.

   (a) Drawing on the distinctions made in this section and Chapter 2, explain the several senses in which someone might claim that how Mr. Kotchian ought to have acted is a "relative" matter. Which of these senses, in your view, would yield true claims and which false?

---

[8] Fergus M. Bordewich, "The Lessons of Bhopal," *Atlantic Monthly,* March 1987, pp. 30–33.
[9] Carl Kotchian, "The Payoff: Lockheed's 70-day Mission to Tokyo," *Saturday Review,* July 9, 1977.

(b) Present and defend your view about whether Mr. Kotchian's actions were morally justified. In doing so, apply utilitarianism, rights ethics, and other ethical theories that you see as relevant.

4. In 1977, the Foreign Corrupt Practices Act was signed into law, largely based on the Lockheed scandal. It makes it a crime for American corporations to accept payments from, or to offer payments to, foreign governments for the purpose of obtaining or retaining business, although it does not forbid "grease" payments to low-level employees of foreign governments (such as clerks) that are part of routine business dealings. Critics are urging repeal of the act because there is no question that it has adversely affected U.S. corporations trying to compete with countries that do not forbid paying business extortion. Is damage to profits a sufficient justification for repealing the act?

5. Since Nigeria became a member of OPEC (the Organization of Oil-Producing Countries) in 1970, the country's oil boom has led to increased corruption, lower living standards for the poor, and much political instability. Foreign oil companies (among whom Shell Oil has received much notoriety) are accused of having disregarded the safety and livelihood of local people when drilling and laying pipelines. The Ogony people in particular have protested, but in vain. Acquaint yourself with the happenings (then and now), and describe what you feel should be the role of foreign oil companies in a country such as Nigeria.

6. The World Trade Organization (WTO) was established to oversee trade agreements, enforce trade rules, and settle disputes. Some troublesome issues have arisen when WTO has denied countries the right to impose environmental restrictions on imports from other countries. Thus, for example, the United States may not impose a ban on fish caught with nets that can endanger other sealife such as turtles or dolphins, while European countries and Japan will not be able to ban imports of beef from U.S. herds injected with antibiotics. Investigate the current disputes and discuss how such problems may be resolved, not overlooking the fact that now a multinational company covering countries A and B has an opportunity to pressure A to relax environmental regulations under the guise of reduced export opportunities to country B, and vice versa regarding exports from B to A. Among other sources, you may wish to consult a contribution by Ralph Nader and Lori Wallach to a book on globalization.[10]

---

[10] Ralph Nader and Lori Wallach, "GATT, NAFTA, and the Subversion of the Democratic Process," ch. 8 in *The Case Against the Global Economy* (San Francisco: Sierra Club Books, 1996).

The Commons and a Livable Environment

Aristotle noted long ago that we tend to be thoughtless about things we do not own individually and that seem to be in unlimited supply. William Foster Lloyd was also an astute observer of this phenomenon. In 1833, he described what the ecologist Garrett Hardin would later call "the tragedy of the commons."[11]

Lloyd observed that cattle in the common pasture of a village were more stunted than those kept on private land. The common fields were themselves more worn than private pastures. His explanation began with the premise that individual farmers are understandably motivated by self-interest to enlarge their herd by one or two cows, especially given that each act taken by itself does negligible damage. Yet the combined effects of all the farmers behaving this way is the (tragic) overgrazing of the pasture so as to damage the good of everyone.

The same kind of competitive, unmalicious, but unthinking exploitation arises with all natural resources held in common: air, land, forests, lakes, oceans, endangered species, and, indeed, the entire biosphere. Hence, the tragedy of the commons remains a powerful image in thinking about environmental challenges in today's era of increasing population and decreasing natural resources. Its very simplicity, however, belies the complexity of many environmental issues. Here we cite several illustrations of that complexity, before turning to various kinds of solutions and philosophical perspectives on the environment.

Cases

Acid Rain—A Surprising Turn of Events

Normal rain has a pH of 5.6, but the typical rain in the northeastern areas of North America is now 3.9 to 4.3. This is 10 to 100 times more acidic than it should be, about as acidic as lemon juice. In addition, the snow-melt each spring releases huge amounts of acid that were in frozen storage during the winter months. Soil that contains natural buffering agents counteracts the acids. But large parts of the northeastern United States and eastern Canada lack sufficient natural buffers to counteract additional onslaughts.

The results? "Acid shock" from snow-melt is thought to have caused annual mass killings of fish. Longer-term effects of the acid harm fish eggs and food sources. Deadly quantities of aluminum, zinc, and many other metals leached from the soil by the acid rain also take a toll as they wash into streams and lakes. It was

---

[11] Garrett Hardin, *Exploring New Ethics for Survival* (New York: Viking, 1968), p. 254.

observed that in the higher elevations of the Adirondack Mountains, more than half the lakes that were once pristine can no longer support fish. Hundreds of other lakes were dying in the United States and Canada. Forests were also dying, and larger animals were suffering dramatic decreases in population, while some farmlands and drinking-water sources were being damaged.

These results occurred during only a few decades. It is believed that North America was just slightly behind Scandinavia, where thousands of lakes have been "killed" by acid rain. In both locations, the cause is now clear: the burning of fossil fuels that release large amounts of sulfur dioxide (SO_2)—the primary culprit—and nitrogen oxides (NOX). In both instances major sources of the pollutants are located hundreds and even thousands of miles away, with winds supplying a deadly transportation system to the damaged ecosystems. Much of Sweden's problem, for example, is traceable to the industrial plants of England and northern Europe. Acid rain problems in Canada and the northeastern United States derive in large measure from the utilities of the Ohio Valley, the largest source of sulfur dioxide pollution in this country. As we know now, pollution does not stop at national borders, necessitating international control of it.

Much remains to be learned about the mechanisms involved in the processes pictured in Figure 6–2. It is still impossible to link specific sources with specific damage. More research into shifting wind patterns and the air transport of acids is needed. Nor is there a reliable estimate of current damage. For example, many believe that microorganisms in soil are being affected in ways that are potentially devastating, but no one knows for sure. Groundwater is undoubtedly being polluted, but it is unclear what that means for human health. Much underground water currently being used was deposited by rainfall over a hundred years ago, and current acid rain may have its main effects on underground water a century from now. Effects on human food sources are also largely unknown. In some areas, certain trees do well; perhaps for them the acid rain acts as a fertilizer.

The good news is that acid rain is now being battled effectively in the United States. When Congress was debating amendments to the Clean Air Act in 1990, industry claimed costs of $3 to $7 billion a year to meet the first stage of reductions by the year 2000, and another $7 to $25 billion a year with the more stringent limits thereafter. But then Congress agreed to let operators of coal burning plants decide on their own how to cut sulphur dioxide emissions: by selecting scrubbers of their own choice or by using costlier cleaner-burning coal, and by *trading* reductions beyond those required. Thus, plants not performing so well could buy emissions credit from the cleaner plants, with dollars changing hands when these plants were not owned by the same company.

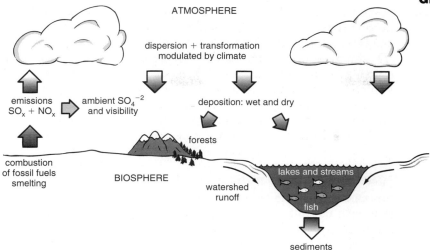

ATMOSPHERE

dispersion + transformation
modulated by climate

emissions
$SO_x + NO_x$

ambient SO_4^{-2}
and visibility

deposition: wet and dry

forests

combustion
of fossil fuels
smelting

BIOSPHERE

watershed
runoff

lakes and streams

fish

sediments

Figure 6–2
Acid deposition: sources and affected ecosystems *(National Research Council,* Acid Deposition, *Long-Term Effects [Washington, DC: National Academy of Sciences, 1986] p. 11. Diagram used with permission of National Academy Press. Also see Donald D. Adams and Walter P. Page, eds.,* Acid Deposition *[New York: Plenum, 1985].)*

All along, the Environmental Protection Agency (EPA) had estimated much lower costs for higher emission reductions, very much to industry's chagrin, but the final results exceeded even EPA's best hopes: by 1996, after the first two years of Phase I limits, emissions dropped to 35 percent *below* the legal limit at a cost of $0.8 billion a year! And EPA's cost estimate for Phase II (years 2000–2010) has dropped to an annual cost of $2.5 billion. It is hoped that this approach, also tried in variations by some European countries, may be useful in also combating greenhouse emissions worldwide.

Other Atmospheric Effects

Other examples can be given of amorphous patterns of ecological damage, like those of acid rain. Worldwide use of fossil fuels by industrial nations is causing a buildup of carbon dioxide in the atmosphere, which could result in a greenhouse effect damaging the entire earth. Some foreseeable effects are changing weather patterns and melting of polar ice-caps leading to raised water lines along ocean coasts. Already observable is an increased bacterial growth rate, as reported by the Physicians for Social Responsibility. Similarly, damage to the protective ozone layer of the earth's atmosphere resulting from the release of chlorofluoro-carbons (CFCs) is related to technological products used by the populations of those same nations. (Where not now prohibited, CFCs such as Freon are used as refrigerants and as propellants in aerosol spray cans.) And rivers amass pollutants as they wind their way through several states or countries, eventually to dump their toxic contents into an ocean. The Rhine is such a river, and the North Sea, now a "special protection area," is such an ocean.

PCBs and Kanemi's Rice Oil

This case of a tainted supply of cooking oil is instructive because even with patients and health officials fully cooperating, a mysterious illness was difficult to define and the offending ingredient extremely hard to trace. How much more effort would it require then to conduct a similar investigation on animals in the wild or in the oceans! Yet their existence has been severely threatened in locations where the same kind of hardy ingredients, polychlorinated biphenyls or PCBs, have been carelessly discharged for decades as industrial waste. A search of web-sites will reveal that this is a world-wide problem and that humans can be affected by eating contaminated fish.

PCBs first appeared as by-products in oil refining processes around the turn of the century and were later recognized to be suitable for use as heat-transfer fluids and as insulating oils in transformers and capacitors. The feathers of stuffed birds in museums indicate that PCB had escaped into the environment before 1914, and from the 1930s to 1950s, producers of PCB oil (e.g., Monsanto) and manufacturers using it (e.g., Westinghouse and GE) were noticing skin rashes and other ailments afflicting workers who handled the oil. The extent of its spread was not recognized until the 1960s when researchers from the coast of Sweden to the coast of California showed that PCB had found its way through the food chain into fish and seabirds and was endangering their natural, biological reproduction. Then, in 1968, following an incident in Japan, the world's public health establishment began to examine PCBs in earnest. Let us pick up the PCB related tragedy in Japan after recalling briefly the sad state of environmental protection there in the 1960s.

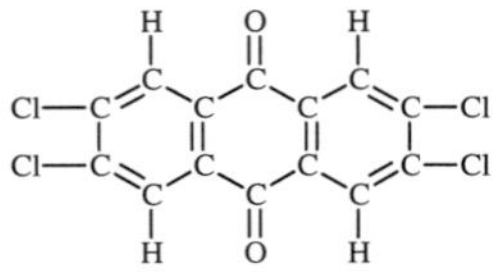

A decade of rapid industrial growth in Japan had taken its toll on the environment. City dwellers fell ill from air pollution. Some rivers were covered with dead fish floating on the surface. The dreadful Minamata disease, from mercury pollution of a nearby bay by the Chisso Company, was continuing to fell its victims (there would be 46 deaths and 75 disabilities by 1970). In May, the itai-itai ("ouch-ouch") disease, which is painful and

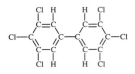

3,3',4,4'-Hexachlorobiphenyl

Figure 6–3
A typical PCB $C_{12}H_4C_{12}$

$C_{12}H_4O_2C_{14}$

2,3,7,8-tetrachlorodibenzo-p-dioxin

Figure 6–4
A typical dioxin, PCB's nasty cousin

causes bone crumbling, was the first illness to be designated by the Japanese government as pollution-generated. Observed on and off since 1920 at various sites, the latest outbreak was placed at the doors of a Mitsui mining and smelting facility that let cadmium escape into a river, from which the pollutant found its way into the food chain via rice paddies and rice. The thalidomide drug tragedy was still on everyone's mind. The severely malformed children of women who had taken that drug during their pregnancies were still not cared for by either the Dainippon Pharmaceutical Company or the government, and schools refused to accept them.[12]

Then, in the summer of 1968, a disease of unknown origin appeared in southern Japan. Victims suffered from disfiguring skin acne and discoloration, fatigue, numbness, respiratory distress, vomiting, and loss of hair. Eventually 10,000 people were stricken and some died. What was the cause? An investigation of 121 cases was conducted, with 121 healthy individuals matched to the victims by age and sex being used as a control group. All 242 were questioned regarding their diets, personal habits, and places of work. When it was discovered that the only significant difference between the two groups was in the amount of fried foods eaten, the disease was traced to rice oil produced by the Kanemi Company.

It took another seven months to find the specific agent in the oil. Autopsies performed on victims revealed the presence of PCBs. Oil made from rice bran at the Kanemi plant was heated at low pressure to remove objectionable odors. The heating pipes were filled with hot Kanechlor, a PCB-containing fluid, but the pipes were corroding and tiny pinholes in them allowed PCBs to leak directly into the oil. In fact, Kanemi had been in the habit of replenishing about 27 kg of lost Kanechlor a month for some time.

There are other, less direct paths by which the extremely toxic PCBs can reach humans. For example, the Kanemi rice oil had also been used as an additive to chicken feed. In early 1968, one million chickens were given this feed and half the chickens died. In the United States, 140,000 chickens were slaughtered in the state of New York on one occasion when data collected by the Campbell's Soup Company revealed more than the permitted level of PCBs in chickens raised by certain growers. The source of PCBs was found to be plastic bakery wrappers mixed in with ground-up stale bread from bakeries used as feed. On another occasion PCBs leaked from a heating system into fishmeal in a

---

[12] Jun Ui, ed., *Polluted Japan* (Tokyo: Jishu-Koza Citizens' Movement, 1972); Nobuko Iijima, *Pollution Japan* (New York: Pergamon, 1979).

North Carolina pasteurization plant. About 12,000 tons of fish-meal were contaminated and 88,000 chickens, already fed this fishmeal, had to be destroyed when the product was recalled.

PCBs were not used only in heat exchangers and electrical equipment. They were also good as hydraulic fluid and plastics additives. But they are no longer considered suitable for such applications, or any other where they can find their way into the environment. Their hardiness accounts for the fact that they were found in the oceans in larger quantities than DDT, although they entered the oceans initially in much lesser amounts. Even a total shutdown of all possible sources of PCB contamination would not result in a rapid reduction in its presence in the environment. Meanwhile production is continuing worldwide. It stood at 57 million pounds a year before 1989. Now that the U.S. has banned PCB applications, the production is still 22 million pounds a year. Information on PCBs is collected by the advocacy group COPA (Coalition Opposed to PCB Ash in Monroe County, Indiana) on its website www.org/library.

Too Little Water for the Everglades

The great marshes of southern Florida have attracted farmers and real estate developers since the beginning of the century. When drained, they present valuable ground. From 1909 to 1912, a fraudulent land development scheme was attempted in collusion with the U.S. Secretary of Agriculture. Arthur Morgan blew the whistle on that situation, jeopardizing not only his own position as a supervising drainage engineer with the U.S. Department of Agriculture, but also that of the head of the Office of Drainage Investigation. An attempt to drain the Everglades was made again by a Florida governor from 1926 to 1929. Once more, Arthur Morgan, this time in private practice, stepped in to reveal the inadequacy of the plans and thus discourage bond sales.

But schemes affecting the Everglades did not end then. Beginning in 1949, the U.S. Army Corps of Engineers started diverting excess water from the giant Lake Okeechobee to the Gulf of Mexico to reduce the danger of flooding to nearby sugar plantations. As a result, the Everglades, lacking water during the dry season, were drying up. A priceless wildlife refuge was falling prey to humanity's appetite. In addition, the diversion of waters to the Gulf and the ocean also affected human habitations in southern Florida. Cities that once thought they had unlimited supplies of fresh groundwater found they were pumping salt water instead as ocean waters seeped in.

Southern Florida is a complex environmental unit with a delicate balance. Any intrusion by human engineering must be seen as an experiment that must be conducted with great care. Unfortunately, too many public agencies view any change in plans as unacceptable once a course has been charted. As Arthur Morgan points out in his book *Dams and Other Disasters,* the Corps was particularly prone to such an attitude, which was fostered by the crisis-oriented training at West Point Military Academy,[13] akin to "group-think." Crises of natural origin (such as floods) may, of course, still befall the Everglades, but then they will gradually recover as they have done by themselves for ages, provided human incursions have not been too severe.

These cases barely hint at the many environmental issues that arise in engineering practice, but they suffice to set a backdrop for distinguishing some ways of addressing environmental and international trade issues. Here we take note of four of these ways, all of which are essential dimensions of workable solutions: industry leadership, governments, market mechanisms, and individuals' commitments.

Corporations: Environmental Leadership

These cases barely hint at the multitude of alarming developments that lead many to speak of an environmental crisis. The good news, however, is that a wide consensus now exists about the importance of environmental issues and the need for concerted action by industry, government, revised market mechanisms, and individuals.

In the present climate, it is simply good business for a corporation to be perceived by the public as environmentally responsible, indeed as a leader in finding creative solutions. Compaq Computer Corporation is only one of a great many encouraging examples.[14] After being founded in 1982, it grew with astonishing success, making the Fortune 500 after only four years. As it did so, it made environmental commitments central to its mission, as recognized with a series of awards, including the 1997 World Environment Center Gold Medal for International Corporate Environmental Achievement.

Three features of Compaq's commitments are especially noteworthy as aspects of its "global" perspective on how its products affected the environment. First, Compaq developed a "life-cycle strategy" for its products that it dubbed "Design for Environment." Priorities were set for efficient use of resources, design of energy-efficient products, easy disassembly for recycling, and waste minimization. For example, it set a timetable for eliminating

---

[13] Arthur E. Morgan, *Dams and Other Disasters* (Boston: Porter Sargent, 1971), pp. 370–89.

[14] Noel M. Tichy, Andrew R. McGill, and Lynda St. Clair, eds., *Corporate Global Citizenship* (San Francisco: New Lexington Press, 1997), pp. 230–44.

CFC emissions in its manufacturing process that was ahead of government requirements, and then met its goal two years ahead of schedule.

Second, Compaq developed unified standards that would apply throughout its operations. This was no minor feat, given that Compaq not only markets its products in over 100 countries, but also has subsidiaries in dozens of countries in North America, Latin America, Europe, the Middle East, Africa, and Asia. Rather than exploiting lower standards in other countries as an excuse to engage in cost savings, Compaq established consistent policies that serve as an exemplar for other companies and industries.

Third, in choosing its suppliers, Compaq places a high priority on companies with a record of environmental concern. Doing so tends to serve its business interests, since some of its costs are shifted to suppliers who already factor in part of the life-cycle concerns. But it also expresses Compaq's genuine and systematic commitments to make environmental responsibility a priority.

Fortunately Compaq is not alone in these efforts. IBM, for instance, also has an extensive (and extensively advertised) computer recycling program. An example of a government run system is Norway's effort to collect computers before they are discarded so they can be refurbished and donated to schools, or they are recycled.

Government: Enabled Natural Disasters and Technology Assessment

Nature's onslaughts can threaten communities and their infrastructures, but the mere occurrence of a hurricane, a tsunami or flood, an earthquake or a landslide, a volcanic eruption or a brush fire, does not in itself constitute a "disaster" or an inevitable "act of God." A disaster is the result of not being prepared for unusual yet predictable natural events. There are four sets of measures communities can take to avert or mitigate disasters without necessarily "stopping" nature.

One set of defensive measures consists of restrictions or requirements imposed on human habitat. For instance, homes should not be built in flood plains, homes in prairie country should have tornado shelters, hillsides should be stabilized to prevent landslides, structures should be able to withstand earthquakes and heavy weather, roof coverings should be made from nonflammable materials, and roof overhangs should be fashioned so flying embers will not be trapped. These are not nightmarish regulations, but merely reminders to developers and builders to do what their profession expects them to do anyway.

Another set of measures consists of strengthening—better yet, duplicating or arranging into grids—the life lines for essential utilities such as water (especially for fire fighting) and electricity. The third category encompasses special purpose defensive structures that would include dams, dikes, breakwaters, avalanche barriers, and means to keep flood waters from damaging low lying sewage plants placed where gravity will take a community's effluents.

A fourth and final set of measures should assure "safe exit" in the form of roads designed as escape routes, structures designated as emergency shelters, adequate clinical facilities, and agreements with neighboring communities for sharing resources in emergencies.

Enabled Natural Disasters

The measures we have cited as examples will not avert emergencies, but they can prevent such emergencies from turning into disasters. Not to take precautions is tantamount to acting as an "enabler" of disaster, much like a person who lacks the foresight or the will to keep alcoholic beverages from a drunkard spouse or partner.

Unfortunately the lessons that could be learned from earlier disasters are often shrugged aside by a disbelief that it could occur again—"Lightning never strikes twice in the same place," and "Another 100-year flood is about that far away,"—or by a belief that government would once more hand out disaster relief payments. For instance, the 1989 Loma Prieta earthquake revealed that the columns of elevated highways needed strengthening, but this has not been fully implemented ten years later. The weaknesses in steel structures found after the 1994 Northridge earthquake have in many cases still not been reported to building departments because the owners claim that any inspection reports they had commissioned should be the owners' private property and concern (and had better not become known to a potential buyer).

Meanwhile the Kobe (Hanshin) earthquake of January 1995 has shown that the infrequency of earthquakes in a particular region is no cause for complacency. In fact, local, regional, and national government agencies were poorly prepared to handle large disasters of any kind in that part of Japan. Even the burning of structural debris after the earthquake was marred by dioxin emanations from the mass of twisted PVC tubing. One outstanding feat, however, was the then almost completed Akashi Kaikyo bridge near Kobe, at 3910 m (with a 1991 m span) the world's longest suspension bridge. The earthquake moved the massive southern anchor and pier by 1 m, so only a few girders had to be modified. The total length is 3911 m now!

On August 16th, 1999 and earthquake hit Izmit and neighboring cities in northwestern Turkey, including Istanbul. The damage to structures and the resulting death toll in the tens of thousands were unusually heavy. And why? Because during a building boom, multistory apartment houses had been built and inspected without serious attention to seismic dangers even though this region was known to be part of an active earthquake belt.

Technology Assessment

What precautionary measures are best? The answers are not easy to find as becomes evident when one considers the many well-intentioned but mishandled large projects of the past, ranging from locating dams or rechanneling rivers to the safe disposal of "spent" nuclear fuel.

Government laws and regulations are understandably the lightning rod in environmental controversies. Few would question the need for the force of law in setting firm guidelines regarding the degradation of the "commons." Yet, how much law, of what sort, and to what ends are matters of continual disagreement.

Until 1995, the U.S. Congress had an Office of Technology Assessment. It prepared studies on the social and environmental effects of technology in areas such as cashless trading (via bank card), nuclear war, health care, and pollution. At the federal and state levels, many large projects must be examined in terms of their environmental impact before they are approved. The purpose of all this activity is praiseworthy. But how effective can it be?

Engineers, it is often said, tend to find the right answers to the wrong questions. The economist Robert Theobald made the following comment on education:

> The university is ideally designed to insure that you remain certain that you know the answers to questions that other people posed long ago. The problem today is that the questions we should be answering are not yet known. Unfortunately the process required for discovering the right questions is totally different from the process of discovering the right answers.[15]

It should be quite apparent that it is not easy to know what questions to ask. And technology assessment and other forecasting methods suffer because of this.

When scientists conduct experiments, they endeavor to distill some key concepts out of their myriad observations. As shown in Figure 6–5, a funnel can be used to portray this activity. At the narrow end of the funnel, we have the current wisdom, the state of the art. Engineers use it to design and build their projects. These develop in many possible directions, as shown by the shape of the lower, inverted funnel. The difficult task of technology assessment and environmental impact analyses is to explore

---

[15] Quoted in Charles A. Thrall and Jerold M. Starr, eds., *Technology, Power, and Social Change* (Lexington, MA: Lexington, 1972), p. 17.

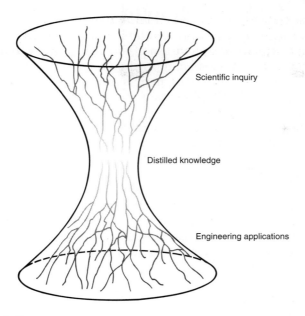

Scientific inquiry

Distilled knowledge

Engineering applications

Figure 6–5
Distilling and applying knowledge.

the extent of this spread and to separate the more significant among the possibly adverse effects.

The danger in any assessment of technology is that some serious risks can easily be overlooked while the studies and subsequent reports, properly authenticated by the aura of scientific methodology, assure the decision maker that nothing is amiss— or perhaps that perceived risks are more serious than they really are. We do not wish to belittle such efforts and we think that they are worthwhile, if only because of those questions they raise— and answers they uncover—that otherwise might not have surfaced. But there is a danger in believing that no further action is required once the reports have been approved and filed. Our contention remains that engineering must be understood as social experimentation and that the experiment continues, indeed enters a new phase, when the engineering project is implemented. Only by careful monitoring, which was well carried out by the Office of Technology Assessment, will it be possible to gather a more complete picture of the tangled web of effects encompassed in Figure 6–5 within the inverted, lower funnel.

Market Mechanisms: Internalizing Costs of Environmental Degradation

Democratic controls can take many forms beyond passing laws. One such option is internalizing costs of harm to the environment. When we are told how efficient and cheap many of our products and processes are—from agriculture to the manufacture of plastics—the figures usually include only the direct costs of labor, raw materials, and use of facilities. If we are quoted a dollar figure, it is at best an approximation of the price. The true cost would have to include numerous indirect factors such as the effects of pollution, the depletion of energy and raw materials, and social costs. If these, or an approximation of them, were internalized (added to the price), then those for whose benefit the environmental degradation had occurred could be charged directly for corrective actions.

As taxpayers are beginning to revolt against higher general levies, the method of having the user of a service or product pay for all its costs is gaining more favor. The engineer must join with the economist, the scientist, the lawyer, and the politician in an effort to find acceptable mechanisms for pricing and releasing products so that the environment is protected through truly self-correcting procedures rather than adequate-appearing yet often circumventable laws.

A working example is the tax imposed by governments in Europe on products and packaging that impose a burden on public garbage disposal or recycling facilities. The manufacturer prepays the tax and certifies so on the product or wrapper.

Fortunately, good design practices may in themselves provide the answers for environmental protection without added real cost. For example, consider the case of a lathe that was redesigned to be vibration-free and manufactured to close tolerances. It not only met occupational safety and health standards for noise, which its predecessor had not, but it also was more reliable and more efficient and had a longer useful life, thus offsetting the additional costs of manufacturing it.[16]

On a more ambitious level, one hopes that the many attempts to produce acceptable and affordable electric and other nonpolluting automobiles which industry is working on will soon succeed.

Individuals: Personal Commitments and Environmental Ethics

Individual engineers can make a difference. Although their actions are limited in a great many ways, they are uniquely placed to act as agents of change, as responsible experimenters. Doing so requires personal commitments based on a combination of moral concern and understanding of wider perspectives. Hence,

---

[16] Seymour Melman, "A Note on: Safety Improvements as a Zero Defect Problem," in *Designing for Safety: Engineering Ethics in Organizational Contexts,* ed. Albert Flores, (Troy, NY: Rensselaer Polytechnic Institute, 1982), p. 176.

we conclude this section with an overview of some of the environmental ethics that are currently being explored by concerned individuals in engineering and other professions.[17]

Human-Centered Environmental Ethics

We have been discussing "environmental ethics" as the study of the moral issues concerning the environment. Now we shift to "environmental ethics" in the sense of a general moral perspective on our moral responsibilities concerning the environment. Each of the ethical theories we have examined provides a foundation for environmental ethics. As human-centered environmental ethics, each focuses on the benefits of the natural environment to humans and the threats to human beings presented by the destruction of nature.

Thus, utilitarians emphasize that human pleasures and interests are linked to nature. Obviously, many of those pleasures and interests are linked to engineered products made from natural resources. In addition, however, we have aesthetic interests, as in the beauty of plants, waterfalls, and mountain ranges; recreational interests, as in hiking and backpacking in wilderness areas; scientific interests, especially in the study of "natural labs" of ecological preserves; and survival interests, which we have learned are linked directly to preserving the natural environment.

Duty ethics urges that respect for human life implies far greater concern for nature than has been traditionally recognized. Kant believed that we owe duties only to rational beings, which in his view excluded all nonhuman animals, although of course he did not have access to recent scientific studies showing striking parallels between humans and other primates. Nevertheless, he condemned callousness and cruelty toward conscious animals because he saw the danger that such attitudes would carry over to indecent treatment of persons. In any case, a duty-centered ethics would emphasize the need for conserving the environment because doing so is implied by respect for human beings who depend on it for their very existence.

---

[17] For helpful sources, see P. Aarne Vesilind and Alastair S. Gunn, *Engineering, Ethics, and the Environment* (New York: Cambridge University Press, 1998); Alastair S. Gunn and P. Aarne Vesilind, eds., *Environmental Ethics for Engineers* (Chelsea, MI: Lewis Publishers, 1986); Joseph R. DesJardins, *Environmental Ethics* 2nd ed. (Belmont, CA: Wadsworth, 1997); Susan J. Armstrong and Richard G. Botzler, eds., *Environmental Ethics* (New York: McGraw-Hill, 1993); Louis P. Pojman, *Environmental Ethics* (Boston: Jones & Bartlett, 1994); Donald VanDeVeer and Christine Pierce, eds., *The Environmental Ethics and Policy Book* (Belmont, CA: Wadsworth, 1994).

Rights ethics urges that the basic right to life entails a right to a livable environment.[18] The right to a livable environment did not generally enter into people's thinking until the end of the twentieth century, at the time when pollution and resource depletion reached alarming proportions. Nevertheless, it is directly implied by the rights to life and liberty, given that these basic rights cannot be exercised without a supportive natural environment.

Finally, virtue ethics draws attention to the virtues of prudence, humility, appreciation of beauty, and gratitude toward the natural world that makes life possible,[19] and also the virtue of stewardship over resources that are needed for further generations.

Nature-Centered Ethics

All these human-centered ethics permit and indeed require a long-term view of conserving the environment. Not everything of importance within a human-centered ethics fits neatly into cost–benefit analyses with limited time horizons; much must be accounted for by means of constraints or limits that cannot necessarily be assigned dollar signs. Yet, some have argued that all versions of human-centered ethics are flawed and that we should widen the circle of things that have inherent worth, that is, value in themselves, independent of human desires and appraisals. Especially since 1979 when the new journal *Environmental Ethics* was founded, philosophers have reported on their explorations of a wide range of *nature-centered ethics* that, for example, affirm the inherent worth of all conscious animals, of all living organisms, or of ecosystems. Religious groups of all faiths have also given voice to their beliefs in human *unity with nature*.

Sentient-Centered Ethics

One version of nature-centered ethics recognizes all sentient animals as having inherent worth. Sentient animals are those that feel pain and pleasure and have desires. Thus, some utilitarians extend their theory (that right action maximizes goodness for all affected) to sentient animals as well as humans. Most notably, Peter Singer developed a utilitarian perspective in his influential book, *Animal Liberation*.[20] Singer insists that moral judgments

---

[18] William T. Blackstone, "Ethics and Ecology," in *Philosophy and the Environmental Crisis,* ed. William T. Blackstone (Athens, GA: University of Georgia Press, 1974).

[19] Thomas E. Hill Jr., "Ideals of Human Excellence and Preserving Natural Environments," *Environmental Ethics* 5 (1983).

[20] Peter Singer, *Animal Liberation,* rev. ed. (New York: Avon Books, 1990).

must take into account the effects of our actions on sentient animals. Failure to do so is a form of discrimination akin to racism and sexism. He labels it "speciesism." Thus, in building a dam that will cause flooding to grasslands, engineers should take into account the impact on animals living there. Singer allows that sometimes animals' interests have to give way to human interests, but their interests should always be considered and weighed.

Singer does not ascribe rights to animals, and hence it is somewhat ironic that *Animal Liberation* has been called the bible of the animal rights movement. Other philosophers, however, do ascribe rights to animals. Most notably, Tom Regan contends that conscious creatures have inherent worth not only because they can feel pleasure and pain, but because, more generally, they are subjects of experiences who form beliefs, memories, intentions, and preferences and can act purposefully.[21] In his view, their status as subjects of experiments makes them sufficiently like humans to give them rights.

Singer and Regan tend to think of inherent worth as all-or-nothing. Hence, they think of conscious animals as deserving equal consideration. That does not mean they must be treated in the identical way we treat humans, but only that their interests should be weighed equally with human interests in making decisions. Other sentient-centered ethicists disagree. They regard conscious animals as having inherent worth, though not equal to that of humans.[22]

Biocentric Ethics

A life-centered ethics regards all living organisms as having inherent worth. Albert Schweitzer (1875–1965) set forth a pioneering version of this perspective under the name of "reverence for life."[23] He argued that the most fundamental feature of us is our will to live, by which he meant both a will to survive and a will to develop according to our innate tendencies. All organisms share these instinctive tendencies to survive and develop, and hence consistency requires that we affirm the inherent worth of all life.

Schweitzer often spoke of reverence for life as the fundamental excellence of character, and hence his view is a version of nature-centered virtue ethics. He refused to rank forms of life

---

[21] Tom Regan, *The Case for Animal Rights* (Berkeley, CA: University of California Press, 1983).

[22] Mary Midgley, *Animals and Why They Matter* (Athens, GA: University of Georgia Press, 1984).

[23] Albert Schweitzer, "The Ethics of Reverence for Life," in *The Philosophy of Civilization,* trans. C. T. Campion (Buffalo, NY: Prometheus Books, 1987), pp. 307–29.

according to degrees of inherent worth, but he believed that a sincere effort to live by the ideal and virtue of reverence for life would enable us to make inevitable decisions about when life must be maintained or has to be sacrificed. More recent defenders of biocentric ethics, however, have developed complex sets of rules for guiding decisions.[24]

Ecocentric Ethics

A frequent criticism of sentient-centered and biocentered ethics is that they are too individualistic, since they locate inherent worth in individual organisms. By contrast, ecocentered ethics locates inherent value in ecological systems. This approach was voiced by the naturalist Aldo Leopold (1887–1948), who urged that we have an obligation to promote the health of ecosystems: "A thing is right when it tends to preserve the integrity, stability, and beauty of the biotic community. It is wrong when it tends otherwise."[25] This "land ethic," as he called it, implied a direct moral imperative to preserve (leave unchanged), not just conserve (use prudently), the environment.

More recent defenders of ecocentric ethics have included within this holistic perspective an appreciation of human relationships. Thus, J. Baird Callicott writes that an ecocentric ethic does not "replace or cancel previous socially generated human-oriented duties—to family and family members, to neighbors and neighborhood, to all human beings and humanity."[26] That is, locating inherent worth in wider ecological systems does not cancel out or make less important what we owe to human beings. This way of thinking is in tune with Native American tribes who live in harmony with nature, based on a sense of respect and reverence for nature, while maintaining a primary respect for one another.

We have set forth these environmental ethics in connection with the reflections of individuals. Clearly, engineering would shut down if it had to grapple with theoretical disputes about human- and nature-centered ethics. Fortunately, at the level of practical issues the ethical theories often converge in the general direction for action, if not in all specifics. Just as humanity is

---

[24] Paul W. Taylor, *Respect for Nature* (Princeton, NJ: Princeton University Press, 1986).

[25] Aldo Leopold, *A Sand County Almanac* (New York: Ballantine, 1970), p. 262.

[26] J. Baird Callicott, "Environmental Ethics," in *Encyclopedia of Ethics,* vol. 1, ed. L. C. Becker (New York: Garland, 1992), pp. 313–14.

part of nature, human-centered and nature-centered ethics overlap extensively in many of their practical implications.[27]

Discussion Topics

1. Exxon's 987-foot tanker *Valdez* was passing through Prince William Sound on March 24, 1989, carrying 50 million gallons of oil when it fetched up on Bligh Reef, tore its bottom, and spilled 11 million gallons of oil at the rate of a thousand gallons a second. This was one of the worst spills ever, not in quantity, but in its effect on a very fragile ecosystem. No human life was lost, but thousands of birds, sea otters, and other creatures died who were stuck in the oil or had fed on the poisoned carcasses.

    Discuss how each of the human-centered and nature-centered ethical theories would interpret the moral issues involved in this case.

2. Consider the following example of environmental side effects cited by Garrett Hardin:

    The Zambesi River . . . was dammed . . . to create the 1700-square-mile Lake Kariba. The effect desired: electricity. The "side-effects" produced: (1) destructive flooding of rich alluvial agricultural land above the dam; (2) uprooting of long-settled farmers from this land to be resettled on poorer hilly land that required farming practices with which they were not familiar; (3) impoverishment of these farmers . . . [and various other social disorders]; (6) creation of a new biotic zone along the lake shore that favored the multiplication of tsetse flies.[28]

    Similar problems have occurred when dams were built in the United States and when the Aswan Dam was erected on the Nile. One might ask if the original purpose may not itself begin to look like merely a side effect. If so, Hardin asks, can we never do anything? Describe under what conditions you think a dam such as the one on the Zambesi River should be built and operated. To whom is the engineer in charge of its construction ultimately responsible?

3. Write an essay on one of the following topics: Why Save Endangered Species?; Why Save the Everglades?; What are Corporations' Responsibilities Concerning the Environment? In your

---

[27] James P. Sterba, "Reconciling Anthropocentric and Nonanthropocentric Environmental Ethics," in *Ethics in Practice,* ed. Hugh LaFollette (New York: Blackwell, 1997), pp. 644–56.

[28] Garrett Hardin, *Exploring New Ethics for Survival* (New York: Viking, 1968), p. 68.

essay discuss the following question: What ethical theory would you apply to our relation to the environment?[29]

4. In part II of *Faust,* Goethe has Faust feverishly planning and supervising a huge land reclamation project. It is fueled by tricky concepts and devices, from paper money to steam engines—note that mephisto remains close at hand—but nature intervenes with tricks of its own. This is how eco-economist Hans Christoff Binswanger comments on Faust's predicament:

> The real danger is that Faust—modern man—will not acknowledge the need for careful planning to forestall such damage as he pushes on relentlessly, not seeing what is going on around him. Goethe symbolizes this blind irresponsibility by Faust's loss of eyesight. . . . Hence mankind compounds its natural limitations—its inability to fully understand nature's complexity—with a blindness induced by hubris.[30]

How do you interpret Goethe's use of the project to reclaim land from the sea?

Weapons Development

> We change our tools and thereafter our tools change us.
> —Marshall McLuhan

Much of the world's technological activity has a military stimulus. Based just on size of expenditures, direct or indirect involvement of engineers, and startling new developments, military technology would deserve serious discussion in these pages. In addition, the moral issues of engineers' participation are of an unparalleled magnitude. They are vitally important even in light of the end of the Cold War. The end of the Cold War has substantially lessened, but not removed, the risk of nuclear war. While this has refocused military concerns on more localized, armed conflicts arising from ethnic/tribal fractiousness, there always looms in the background the urge of nations to secure—if necessary, by force—adequate supplies of water, energy, and other vital resources for their growing populations.

---

[29] See Nicholas Rescher, "Why Save Endangered Species?" in *Unpopular Essays and Technological Progress* (Pittsburgh: University of Pittsburgh Press, 1980), pp. 79–92; Bryan G. Norton, *Why Preserve Natural Variety?* (Princeton, NJ: Princeton University Press, 1987).

[30] Hans Christoff Binswanger, "The Challenge of Faust," SCIENCE, v. 281, 31 July 1998, pp. 640–641. See also Binswanger's book *Money and Magic: A Critique of the Modern Economy in Light of Goethe's Faust* (University of Chicago Press, 194).

The 1991 Gulf War and the 1999 Kosovo action made clear the dual dangers of military confrontation and temptations to engage in high-tech wars. The continuing automation of the battle scene tends to conceal the real horrors of war and thus makes military activity seem less threatening and nonmilitary resolutions less pressing.

> Technology is
> neither good,
> nor bad, nor is it neutral.
> —Marvin Kranzberg

The Engineer's Involvement in Weapons Work

Instruments of torture leave little to the imagination. Descriptions of the use of the rack, the thumbscrew, and the electric prod convey instant sensations of pain and suffering. Historically, a quick death in battle by sword was considered acceptable, while the use of remote weapons (from bow and arrow to firearms) was frequently decried as cowardly, devoid of valor, and tantamount to plain murder.[31]

As our modern weapons of war have progressed through catapults, cannons, machine guns, and bombs released from airplanes and missiles to reach further and further, even the soldiers firing them are less likely to see the individual human beings—soldiers as well as civilians—they had as their general target. On the other hand, a survey of U.S. front line soldiers of World War II revealed that when they had already aimed their guns at individually discernible enemy soldiers within range, more than half did *not* pull the trigger (and not because they were afraid to reveal their own positions).

How do the men and women who design, manufacture, and use weapons feel about their work? Many have reservations about their tasks, but those who stick it out also have reasons to support their continued involvement. The following cases involve real weapons. The people are composites who represent the various positions taken by typical citizens who are also engineers.

1. Bob's employer manufactures antipersonnel bombs. By clustering 665 guava-sized bomblets and letting them explode above

---

[31] Martin von Creveld, *Technology and War, from 2000 B.C. to the Present* (New York and London: The Free Press/Macmillan, 1989), p. 71.

ground, an area covering the equivalent of 10 football fields is subjected to a shower of sharp fragments. Alternatively, the bombs can be timed to explode hours apart after delivery. Originally the fragments were made of steel, thus often removable with magnets; now plastic materials are sometimes used, making the treatment of wounds, including the location and removal of the fragments, more time-consuming for the surgeon. Recently, another innovation was introduced: by coating the bomblets with phosphorus, the fragments could inflict internal burns as well. Thus, the antipersonnel bomb does its job quite well without necessarily killing in that it ties up much of the enemy's resources just by treating the wounded who have survived.

Bob himself does not handle the bombs in any way, but as an industrial engineer, he enables the factory to run efficiently. He does not like to be involved in making weapons, but then he tells himself that someone has to produce them. If he does not do his job, someone else will, so nothing would change. Furthermore, with the cost of living being what it is, he owes his family a steady income.

2. Mary is a chemical engineer. A promotion has gotten her into napalm manufacturing. She knows it is nasty stuff. She remembers Professor Wald, a Nobel laureate in biology from Harvard, berating the chemical industry for producing this "most brutal and destructive weapon that has ever been created." But this was when she was in college, during the Vietnam war. Civilians were forever not leaving the fighting zone and then there were complaints about them being hurt or killed. She abhors war like most human beings, but she feels that the government knows more than she does about international dangers and that it is better to fight a war abroad than on our shores. If everyone were to decide on her or his own what to do and what not to do, then there would be utter chaos. Perhaps society can tolerate a few oddballs with their own ideas, but companies certainly should be prepared to manufacture the weapons our armed forces need. Incidentally, if Mary continues to perform well on her job, she will be promoted out of her present position into working on a commercial product with much growth potential.

3. Ron is a specialist in missile control and guidance. He is proud to be able to help his country through his efforts in the defense industry. The missiles on which he works will carry single or multiple warheads with the kind of dreadful firepower that, in his estimation, has kept any potential enemy in check since 1945. At least there has not been another world war—the result of mutual deterrence, he believes.

4. Marco's foremost love is physical electronics. He works in one of the finest laser laboratories. Some of his colleagues do excit-

ing research in particle beams. That the laboratory is interested in developing something akin to the "death ray" described by science fiction writers of his youth is of secondary importance. More bothersome is the secrecy that prevents him from freely exchanging ideas with experts across the world. But why change jobs if he will never find facilities like those he has now?

5. Joanne is an electronics engineer whose work assignment includes avionics for fighter planes that are mostly sold abroad. She has no qualms about such planes going to what she considers friendly countries, but she draws the line at their sale to potentially hostile nations. Joanne realizes that she has no leverage within the company, so she occasionally alerts journalist friends with news she feels all citizens should have. "Let the voters direct the country at election time"—that is her motto.

6. Ted's background and advanced degrees in engineering physics gave him a ready entry into nuclear bomb development. As a well-informed citizen, he is seriously concerned with the dangers of the ever-growing nuclear arsenal. He is also aware of the possibilities of an accidental nuclear exchange. In the meantime, he is working hard to reduce the risk of accidents such as the 32 "broken arrows" (incidents when missile launchings may have occurred erroneously) that had been reported by the Pentagon during the height of the Cold War, or the many others that he knows have occurred worldwide. Ted continues in his work because he believes that only specialists, with first-hand experience of what modern weapons can do, can eventually turn around the suicidal trend represented by their development. Who else can engage in meaningful arms control negotiations?

The names are fictitious, as are the specific jobs described, but they suffice to illustrate the kinds of moral issues involved in deciding whether to engage in military work. Prudential self-interest is not sufficient to guarantee responsible participation in what must be regarded as humankind's most crucial engineering experiment. We must rely on individuals who have arrived at morally autonomous, well-reasoned positions for either engaging in or abstaining from weapons work to consistently and carefully monitor the experiment and try to keep it from running a wild course.

For someone with an inventive mind, it may be difficult to comprehend how others may look at the effect of some inventions. "On April 13, 1888, when Alfred A. Nobel—a Swedish chemist living in Paris—awoke to read the morning paper, he found his own obituary. Mistakenly run in place of one for his older brother, Ludwig (who'd died the day before in Russia), the item left Nobel stunned."[32] Why? Because in strong words, the obituary described Nobel as being responsible for the frantic arms race gripping Europe. Nobel saw himself in a different role: he had invented means to make explosives safe to use. After five months in depression, he renounced all explosive experiments and many of his 350 patents that had made him very wealthy, then established the Nobel Prizes.

Defense Industry Problems

Nations confer special privileges on their defense industries, often without giving sufficient thought to the problems that can accompany large military buildups. Unethical business practices, for instance, occur as in all massive projects, but the urgency of completing a weapons system before it becomes obsolete and the secrecy that surrounds it make proper oversight particularly difficult. This is one of the problems we describe briefly below. The other problems are even more serious because they frequently go unrecognized.

The problem of waste and cost overruns is a continuing one in the defense industry.[33] One example (earlier discussed as a whistleblower case in Chapter 5) is the $2 billion cost overrun on development of the C5-A transport plane, revealed to the U.S. Senate by Ernest Fitzgerald, a Pentagon specialist in management systems. Fitzgerald has been a critic of how the defense industry has operated at efficiencies far below commercial standards. He has described how contractors' workforces were swelled with underutilized engineers and high-salary sales personnel, resulting in lavish overhead fees. Or how small contractors were willing to comply with cost-cutting plans, but large suppliers felt secure in not complying.

High cost and poor quality resulted from various practices: Planned funding levels were leaked to prospective contractors. Cost estimates were based on historical data, thus incorporating past inefficiencies. Costs were cut when necessary by lowering quality, especially when component specifications were not ready until the contract was completed. Sole-supplier policies gave a contractor the incentive to "buy in" with an artificially low bid, only to plead for additional funds later on. And those funds were usually forthcoming, since the Department of Defense has historically accepted what it knows to be optimistically low

---

[32] Donovan Webster, *Aftermath: The Remnants of War* (New York: Vintage Books/Random House, 1996).
[33] Ernest Fitzgerald, *The High Priests of Waste* (New York: W. W. Norton, 1972); J. Gansler, *The Defense Industry* (Cambridge, MA: MIT Press, 1980); Seymour Melman, *Pentagon Capitalism* (New York: McGraw-Hill, 1970).

development-cost estimates because they stand a better chance of being approved by Congress.[34]

In Goethe's poem *Der Zauberlehrling,* the sorcerer's apprentice employs his master's magic incantation to make the broom fetch water. When he cannot remember the proper command to stop the helpful broom, however, he comes near to drowning before the master returns. Military technology often resembles the sorcerer's broom. Not only has the world's arsenal grown inordinately expensive (even without graft), and not only does it contribute to a steadily worsening inflation, it has also gained a momentum all its own. And along with it the temptation to test new weapons in practice, even when the manufacturer's country is not under attack. Occasions presented themselves during the Spanish Civil War, and on several occasions after World War II.

From the 1960s to the late 1980s, the arsenal grew in sophistication as well as size. Diplomats may be striving to avert military conflicts, but all the while an exuberance for new developments creates a technology creep that can at times postpone or even upset all negotiations. Nations are suddenly seen to shift to new positions as new devices to more accurately target missiles, or perhaps an entirely new weapon, are reported to be in the offing. This can destabilize (and occasionally stabilize) the political process.

The technological imperative that innovations must be implemented should even give advocates of preparedness for conventional, limited war some cause for concern. Giving in to the excitement of equipping and trying out weapons employing the latest in technology may provide added capability to sophisticated, fully automatic systems such as intercontinental ballistic missiles. But if tactical, humanly operated weapons fall prey to the gadget craze, a less-than-optimal system may result. The F-15 fighter illustrates this problem of preoccupation with prestige-boosting modernism. The plane was the fastest and most maneuverable of its kind, yet 40 percent of the F-15s were not available for service at any one time because of defects, difficulty of repair, and lack of spare parts.

A further problem concerns peacetime secrecy in work of military import. Secrecy poses problems for engineers in various ways. Should discoveries of military significance always be made available to the government? Can they be shared with other researchers, with other countries? Or should they be withheld from the larger scientific and public community altogether? If governmental secrecy in weapons development is allowed to become

---

[34] Gansler, *The Defense Industry,* p. 296.

all-pervasive, on the other hand, will it also serve to mask corruption or embarrassing mistakes within the defense establishment? Can secrecy contribute to the promotion of particular weapon systems, such as the x-ray laser, without fear of criticism?[35] There are no easy answers to these questions, and they deserve to be discussed more widely within defense establishments and in public.

Decommissioning Weapons and Lasting Effects

To this day, farmers in France plow up shells, duds or live, that landed in the ground more than $^3/_4$ of a century ago during World War I. Over time, the repetitive cycle of freeze and thaw has caused these shells to move close to the surface. Special bomb disposal units keep busy with hundreds of calls a year, and injuries occur as well. To this must be added the many still-hidden, unexploded bombs that fell all over the world during World War II.

More recent is the grim harvest of severed limbs and dead bodies as peasants and their treasured water buffaloes tread on mines planted by all sides in Cambodia and Vietnam in the 1960s and 1970s. Afghanistan, Angola, Bosnia, Mozambique, Nicaragua, Somalia, and the Balkans are other regions infested by such antipersonnel weapons. They are easily spread by air but painstakingly difficult and dangerous to remove. The U.S. State Department estimates that 85 to 100 million land mines still remain scattered in the countries listed above and those that were involved in the two world wars.

Clearly, the design, manufacture, and deployment of weapons of all kinds is a huge experiment that includes not only their use but also eventual disposal of the arsenal by means other than export. In terms of pure experimentation, the real tragedy of widespread ignorance regarding radiation is only now being revealed. Experiments with soldiers observing atomic bomb blasts, patients being treated with radiation as guinea pigs, and radium inserted for the "modern" treatment of persistent nasal and throat problems—only to induce brain cancer in many cases years later—are just a few of the poorly conceived measures during and after World War II when radiation was both an officially imposed secret among the knowledgeable and a cure-all for others. The use of agent orange defoliants in Vietnam is only now officially recognized as a health hazard as U.S. soldiers show symptoms of ill effects, long after scientists warned of its effects on farmers and their animals in the war zones of Vietnam. Gas

---

[35] John A. Adam, "Dispute over X-Ray Laser Made Public, Scientists' Criticisms Shunted," *The Institute,* February 1988, p. 1.

warfare experiments had their share of involuntary subjects among soldiers of the United States and Australia. In the former Soviet Union, anthrax carriers were released accidentally from biological warfare plants and affected tens of thousands of people living downwind. The ultimate horror, of course, will always be the intentional gassing and human experimentation in Nazi concentration camps. But engineers and scientists dealing with materials dangerous enough to be considered useful as weapons must consider not only their direct use but also their accidental diversion and ultimate safe disposal. After all, structural engineers are not free to build huge structures without considering how to safely dismantle them eventually.

Discussion Topics

1. The following problem is taken from an article by Tekla Perry in the *IEEE Spectrum.* Although it involves the National Aeronautics and Space Administration rather than the Defense Department, many of the actors (companies and government) involved in space research are also involved in weapons development:

   Arthur is chief engineer in a components house. As such, he sits in meetings concerning bidding on contracts. At one such meeting between top company executives and the National Aeronautics and Space Administration, which is interested in getting a major contract, NASA presents specifications for components that are to be several orders of magnitude more reliable than the current state of the art. The components are not part of a life-support system, yet are critical for the success of several planned experiments. Arthur does not believe such reliability can be achieved by his company or any other, and he knows the executives feel the same. Nevertheless, the executives indicate an interest to bid on the contract without questioning the specifications. Arthur discusses the matter privately with the executives and recommends that they review the seemingly technical impossibility with NASA and try to amend the contract. The executives say that they intend, if they win the contract, to argue mid-stream for a change. They remind Arthur that if they don't win the contract, several engineers in Arthur's division will have to be laid off. Arthur is well-liked by his employees and fears the lay-offs would affect some close friendships. What should Arthur do?[36]

---

[36] Tekla S. Perry, "Five Ethical Dilemmas," *IEEE Spectrum* 18 (June 1981), p. 58. Quotations in text used with permission of the author and the Institute of Electrical and Electronics Engineers.

2. Are there any ethical grounds for maintaining a large nuclear stockpile today? Discuss any stabilizing or destabilizing effects you see. Also, discuss the promise and perils of the "swords-to-plowshares" idea of selling weapons-grade plutonium to utility companies for use in nuclear power plants.[37]

3. The just-war theory considers a war acceptable when it satisfies several stringent criteria: The war must be fought for a just cause, the motives must be good, it must follow a call from higher authority to legitimize it, and the use of force must be based on necessity.[38] Central to notions of a just war are the principles of noncombatant immunity and proportionality. Noncombatants are those who will not be actively participating in combat and therefore do not need to be killed or restrained. Proportionality addresses the extent of damage or consequences allowable in terms of need and cost. Describe a scenario for the conduct of a just war and describe the kinds of weapons engineers might have to develop to wage one.

4. Inform yourself on what happens to unexploded bombs and mines after the end of a war.[39] Should munitions that deactivate themselves with age be substituted?

---

[37] See U.S. Catholic Bishops, "On the Use of Nuclear Weapons and Nuclear Deterrence," from *The Challenge of Peace: God's Promise and Our Response* (Washington, DC: United States Catholic Conference, 1982); Gregory S. Kavka, *Moral Paradoxes of Nuclear Deterrence* (Cambridge: Cambridge University Press, 1987); James P. Sterba, *The Ethics of War and Nuclear Deterrence* (Belmont, CA: Wadsworth, 1985); Frank Clifford, "Plutonium from Bombs May Be Used in Reactors," *Los Angeles Times,* November 15, 1998, p. 1.

[38] J. Bryan Hehir, "The Relationship of Moral and Strategic Arguments in the Defense Debate," in *Research in Philosophy and Technology,* vol. 3, ed. Paul T. Durbin (Greenwich, CT: JAI Press, 1980), pp. 368–69.

[39] Webster, *Aftermath: The Remnants of War.*

Appendix

Appendix A1: A Taxonomy of Failures in Engineering

This table lists common examples of the types of failures that occur in the practice of engineering. The first column indicates the activity that is the source of a failure, and the next three columns indicate who may be affected by the activity. The activities are divided into three groups:

1) Engineering related tasks: Those undertaken by engineers and related professionals;
2) Computer related tasks: those undertaken by computer scientists, engineers, and programmers;
3) Business related tasks: those undertaken by engineers and others serving as managers, purchasing agents, and sales personnel.

The numbers in the body of the Table refer to cases selected from the larger number of cases (and not only failures) given in the text. The cases are identified by these numbers in the adjacent List of Cases (Appendix A2).

TABLE A-1

TASK	PARTIES AFFECTED BY THE TASK		
	WORKERS in Manufacturing and Construction	**OPERATORS** and USERS of the Product	**PUBLIC** and the NATURAL ENVIRONMENT
1) Engineering Tasks			
Designing a product (conceptual & final)	44	7, 8, 9, 15, 32, 35, 41, 49, 50, 58	3, 4, 9, 14, 21, 27, 30, 39, 42, 50
Producing it (in a shop or at the construction site)	6	6, 24, 37	29, 33, 42, 45
Operating the product in system	10, 24	2, 8, 25, 40, 41, 47, 52	1, 2, 3, 4, 9, 10, 19, 20, 44, 46
Maintaining and dismantling the product	10, 35	35, 50	26, 29, 34, 50
2) Computing Tasks			
Producing and operating computer programs for storing, searching, and processing data		54	18, 54
Controlling communications and machines		34	34, 38, 42
3) Business Tasks		3, 37, 40, 43, 49	5, 28, 48
Sales		5	3, 22, 37
Purchasing		22	22

Appendix A2: Index of Selected, Illustrative Cases

A selection of cases mentioned in this book are listed below by keywords in alphabetical order. Each one has the number by which it is listed in the Taxonomy table (Appendix A1) and a page number indicating where the case is mentioned in the text.

Appendix A3: Code of Ethics

Codes of some engineering societies of an interdisciplinary nature:

1. ABET Accreditation Board for Engineering and Technology (p. 226)
2. NSPE National Society of Professional Engineers (p. 227)

Codes of some discipline-specific engineering societies:

3. AIChE American Institute of Chemical Engineers (p. 235)
4. ASCE American Society of Civil Engineers (p. 236)
5. ASME American Society of Mechanical Engineers (p. 241)
6. IEEE Institute of Electrical and Electronics Engineers (p. 246)

The first code adopted by an engineering society in the United States:

7. AIEE American Institute of Electrical Engineering, a parent of IEEE (p. 247)

Historical Note:

The AIEE was the first of the five so-called founder engineering societies in the U.S. to adopt an engineering code. That code was adopted in 1912 (see page 247). In 1963 the AIEE merged with the younger but larger Institute of Radio Engineers to create the Institute of Electrical and Electronics Engineers (IEEE), now an organization of international scope. The present IEEE code is listed here in alphabetical order with the codes of three other founder societies (AIChE, ASCE, ASME). Sources for codes of other societies include: Center for Studies of Ethics in the Professions, Illinois Inst. Tech., http://csep.iit.edu/codes; The Online Ethics Center for Engineering and Science, http://onlineethics.org/codes/codes.html; World Codes, Virginia Inst. of Technology, course material, http://ei.cs.vt.edu/~cs3604/lib, click on WorldCodes.

**Accreditation
Board for
Engineering and
Technology
(ABET)***

The Fundamental Principles

Engineers uphold and advance the integrity, honor and dignity of the engineering profession by:

  I. using their knowledge and skill for the enhancement of human welfare;
 II. being honest and impartial, and serving with fidelity the public, their employers and clients;

---

* 1977 version. Formerly Engineers' Council for Professional Development. Reprinted with permission of ABET.

III. striving to increase the competence and prestige of the engineering profession; and

IV. supporting the professional and technical societies of their disciplines.

The Fundamental Canons

1. Engineers shall hold paramount the safety, health and welfare of the public in the performance of their professional duties.
2. Engineers shall perform services only in the areas of their competence.
3. Engineers shall issue public statements only in an objective and truthful manner.
4. Engineers shall act in professional matters for each employer or client as faithful agents or trustees, and shall avoid conflicts of interest.
5. Engineers shall build their professional reputation on the merit of their services and shall not compete unfairly with others.
6. Engineers shall act in such a manner as to uphold and enhance the honor, integrity and dignity of the profession.
7. Engineers shall continue their professional development throughout their careers and shall provide opportunities for the professional development of those engineers under their supervision.

Preamble

National Society of Professional Engineers (NSPE)*

Engineering is an important and learned profession. The members of the profession recognize that their work has a direct and vital impact on the quality of life for all people. Accordingly, the services provided by engineers require honesty, impartiality, fairness and equity, and must be dedicated to the protection of the public health, safety and welfare. In the practice of their profession, engineers must perform under a standard of professional behavior which requires adherence to the highest principles of ethical conduct on behalf of the public, clients, employers and the profession.

I. Fundamental Canons
Engineers, in the fulfillment of their professional duties, shall:
1. Hold paramount the safety, health and welfare of the public in the performance of their professional duties.

* 1993 version. Reprinted with permission of the NSPE.

2. Perform services only in areas of their competence.
3. Issue public statements only in an objective and truthful manner.
4. Act in professional matters for each employer or client as faithful agents or trustees.
5. Avoid deceptive acts in the solicitation of professional employment.

II. Rules of Practice
1. Engineers shall hold paramount the safety, health and welfare of the public in the performance of their professional duties.
   a. Engineers shall at all times recognize that their primary obligation is to protect the safety, health, property and welfare of the public. If their professional judgment is overruled under circumstances where the safety, health, property or welfare of the public are endangered, they shall notify their employer or client and such other authority as may be appropriate.
   b. Engineers shall approve only those engineering documents which are safe for public health, property and welfare in conformity with accepted standards.
   c. Engineers shall not reveal facts, data or information obtained in a professional capacity without the prior consent of the client or employer except as authorized or required by law or this Code.
   d. Engineers shall not permit the use of their name or firm name nor associate in business ventures with any person or firm which they have reason to believe is engaging in fraudulent or dishonest business or professional practices.
   e. Engineers having knowledge of any alleged violation of this Code shall cooperate with the proper authorities in furnishing such information or assistance as may be required.
2. Engineers shall perform services only in the areas of their competence.
   a. Engineers shall undertake assignments only when qualified by education or experience in the specific technical fields involved.
   b. Engineers shall not affix their signatures to any plans or documents dealing with subject matter in which they lack competence, nor to any plan or document not prepared under their direction and control.
   c. Engineers may accept assignments and assume responsibility for coordination of an entire project and sign and seal the engineering documents for the

entire project, provided that each technical segment is signed and sealed only by the qualified engineers who prepared the segment.

3. Engineers shall issue public statements only in an objective and truthful manner.

   a. Engineers shall be objective and truthful in professional reports, statements or testimony. They shall include all relevant and pertinent information in such reports, statements or testimony.

   b. Engineers may express publicly a professional opinion on technical subjects only when that opinion is founded upon adequate knowledge of the facts and competence in the subject matter.

   c. Engineers shall issue no statements, criticisms or arguments on technical matters which are inspired or paid for by interested parties, unless they have prefaced their comments by explicitly identifying the interested parties on whose behalf they are speaking, and by revealing the existence of any interest the engineers may have in the matters.

4. Engineers shall act in professional matters for each employer or client as faithful agents or trustees.

   a. Engineers shall disclose all known or potential conflicts of interest to their employers or clients by promptly informing them of any business association, interest, or other circumstances which could influence or appear to influence their judgment or the quality of their services.

   b. Engineers shall not accept compensation, financial or otherwise, from more than one party for services on the same project, or for services pertaining to the same project, unless the circumstances are fully disclosed to, and agreed to by, all interested parties.

   c. Engineers shall not solicit or accept financial or other valuable consideration directly or indirectly, from contractors, their agents, or other parties in connection with work for employers or clients for which they are responsible.

   d. Engineers in public service as members, advisors or employees of a governmental or quasi-governmental body or department shall not participate in decisions with respect to professional services solicited or provided by them or their organizations in private or public engineering practice.

   e. Engineers shall not solicit or accept a professional contract from a governmental body on which a principal or officer of their organization serves as a member.

5. Engineers shall avoid deceptive acts in the solicitation of professional employment.

   a. Engineers shall not falsify or permit misrepresentation of their, or their associates', academic or professional qualifications. They shall not misrepresent or exaggerate their degree of responsibility in or for the subject matter of prior assignments. Brochures or other presentations incident to the solicitation of employment shall not misrepresent pertinent facts concerning employers, employees, associates, joint ventures or past accomplishments with the intent and purpose of enhancing their qualifications and their work.

   b. Engineers shall not offer, give, solicit or receive, either directly or indirectly, any political contribution in an amount intended to influence the award of a contract by public authority, or which may be reasonably construed by the public of having the effect or intent to influence the award of a contract. They shall not offer any gift, or other valuable consideration in order to secure work. They shall not pay a commission, percentage or brokerage fee in order to secure work except to a bona fide employee or bona fide established commercial or marketing agencies retained by them.

III. Professional Obligations

   1. Engineers shall be guided in all their professional relations by the highest standards of integrity.

      a. Engineers shall admit and accept their own errors when proven wrong and refrain from distorting or altering the facts in an attempt to justify their decisions.

      b. Engineers shall advise their clients or employers when they believe a project will not be successful.

      c. Engineers shall not accept outside employment to the detriment of their regular work or interest. Before accepting any outside employment they will notify their employers.

      d. Engineers shall not attempt to attract an engineer from another employer by false or misleading pretenses.

      e. Engineers shall not actively participate in strikes, picket lines, or other collective coercive action.

      f. Engineers shall avoid any act tending to promote their own interest at the expense of the dignity and integrity of the profession.

2. Engineers shall at all times strive to serve the public interest.

    a. Engineers shall seek opportunities to be of constructive service in civic affairs and work for the advancement of the safety, health and well-being of their community.

    b. Engineers shall not complete, sign or seal plans and/or specifications that are not of a design safe to the public health and welfare and in conformity with accepted engineering standards. If the client or employer insists on such unprofessional conduct, they shall notify the proper authorities and withdraw from further service on the project.

    c. Engineers shall endeavor to extend public knowledge and appreciation of engineering and its achievements and to protect the engineering profession from misrepresentation and misunderstanding.

3. Engineers shall avoid all conduct or practice which is likely to discredit the profession or deceive the public.

    a. Engineers shall avoid the use of statements containing a material misrepresentation of fact or omitting a material fact necessary to keep statements from being misleading or intended or likely to create an unjustified expectation, or statements containing prediction of future success.

    b. Consistent with the foregoing, Engineers may advertise for recruitment of personnel.

    c. Consistent with the foregoing, Engineers may prepare articles for the lay or technical press, but such articles shall not imply credit to the author for work performed by others.

4. Engineers shall not disclose confidential information concerning the business affairs or technical processes of any present or former client or employer without his consent.

    a. Engineers in the employ of others shall not without the consent of all interested parties enter promotional efforts or negotiations for work or make arrangements for other employment as a principal or to practice in connection with a specific project for which the Engineer has gained particular and specialized knowledge.

    b. Engineers shall not, without the consent of all interested parties, participate in or represent an adversary interest in connection with a specific project or proceeding in which the Engineer has gained particular specialized knowledge on behalf of a former client or employer.

5. Engineers shall not be influenced in their professional duties by conflicting interests.
   a. Engineers shall not accept financial or other considerations, including free engineering designs, from material or equipment suppliers for specifying their product.
   b. Engineers shall not accept commissions or allowances, directly or indirectly, from contractors or other parties dealing with clients or employers of the Engineer in connection with work for which the Engineer is responsible.
6. Engineers shall uphold the principle of appropriate and adequate compensation for those engaged in engineering work.
   a. Engineers shall not accept remuneration from either an employee or employment agency for giving employment.
   b. Engineers, when employing other engineers, shall offer a salary according to professional qualifications.
7. Engineers shall not attempt to obtain employment or advancement or professional engagements by untruthfully criticizing other engineers, or by other improper or questionable methods.
   a. Engineers shall not request, propose, or accept a professional commission on a contingent basis under circumstances in which their professional judgment may be compromised.
   b. Engineers in salaried positions shall accept part-time engineering work only to the extent consistent with policies of the employer and in accordance with ethical considerations.
   c. Engineers shall not use equipment, supplies, laboratory, or office facilities of an employer to carry on outside private practice without consent.
8. Engineers shall not attempt to injure, maliciously or falsely, directly or indirectly, the professional reputation, prospects, practice or employment of other engineers, nor untruthfully criticize other engineers' work. Engineers who believe others are guilty of unethical or illegal practice shall present such information to the proper authority for action.
   a. Engineers in private practice shall not review the work of another engineer for the same client, except with the knowledge of such engineer, or unless the connection of such engineer with the work has been terminated.
   b. Engineers in governmental, industrial or educational employ are entitled to review and evaluate the work

of other engineers when so required by their employment duties.

 c. Engineers in sales or industrial employ are entitled to make engineering comparisons of represented products with products of other suppliers.

9. Engineers shall accept personal responsibility for their professional activities; provided, however, that Engineers may seek indemnification for professional services arising out of their practice for other than gross negligence, where the Engineer's interests cannot otherwise be protected.

 a. Engineers shall conform with state registration laws in the practice of engineering.

 b. Engineers shall not use association with the non-engineer, a corporation, or partnership as a "cloak" for unethical acts, but must accept personal responsibility for all professional acts.

10. Engineers shall give credit for engineering work to those to whom credit is due, and will recognize the proprietary interests of others.

 a. Engineers shall, whenever possible, name the person or persons who may be individually responsible for designs, inventions, writings, or other accomplishments.

 b. Engineers using designs supplied by a client recognize that the designs remain the property of the client and may not be duplicated by the Engineer for others without express permission.

 c. Engineers, before undertaking work for others in connection with which the Engineer may make improvements, plans, designs, inventions, or other records which may justify copyrights or patents, should enter into a positive agreement regarding ownership.

 d. Engineers' designs, data, records, and notes referring exclusively to an employer's work are the employer's property.

11. Engineers shall cooperate in extending the effectiveness of the profession by interchanging information and experience with other engineers and students, and will endeavor to provide opportunity for the professional development and advancement of engineers under their supervision.

 a. Engineers shall encourage engineering employees' efforts to improve their education.

 b. Engineers shall encourage engineering employees to attend and present papers at professional and technical society meetings.

c. Engineers shall urge engineering employees to become registered at the earliest possible date.

d. Engineers shall assign a professional engineer duties of a nature to utilize full training and experience, insofar as possible, and delegate lesser functions to subprofessionals or to technicians.

e. Engineers shall provide a prospective engineering employee with complete information on working conditions and proposed status of employment, and after employment will keep employees informed of any changes.

"By order of the United States District Court for the District of Columbia, former Section 11(c) of the NSPE Code of Ethics prohibiting competitive bidding, and all policy statements, opinions, rulings or other guidelines interpreting its scope, have been rescinded as unlawfully interfering with the legal right of engineers, protected under the antitrust laws, to provide price information to prospective clients; accordingly, nothing contained in the NSPE Code of Ethics, policy statements, opinions, rulings or other guidelines prohibits the submission of price quotations or competitive bids for engineering services at any time or in any amount."

Statement by NSPE Executive Committee

In order to correct misunderstandings which have been indicated in some instances since the issuance of the Supreme Court decision and the entry of the Final Judgment, it is noted that in its decision of April 25, 1978, the Supreme Court of the United States declared: "The Sherman Act does not require competitive bidding."

It is further noted that as made clear in the Supreme Court decision:

1. Engineers and firms may individually refuse to bid for engineering services.

2. Clients are not required to seek bids for engineering services.

3. Federal, state, and local laws governing procedures to procure engineering services are not affected, and remain in full force and effect.

4. State societies and local chapters are free to actively and aggressively seek legislation for professional selection and negotiation procedures by public agencies.

5. State registration board rules of professional conduct, including rules prohibiting competitive bidding for engineering services, are not affected and remain in full force and effect. State registration boards with authority to adopt rules of professional conduct may adopt rules governing procedures to obtain engineering services.

6. As noted by the Supreme Court, "nothing in the judgment prevents NSPE and its members from attempting to influence governmental action. . . ."

Note: In regard to the question of application of the Code to corporations vis-a-vis real persons, business form or type should not negate nor influence conformance of individuals to the Code. The Code deals with professional services, which services must be performed by real persons. Real persons in turn establish and implement policies within business structures. The Code is clearly written to apply to the Engineer and it is incumbent on a member of NSPE to endeavor to live up to its provisions. This applies to all pertinent sections of the Code.

Publication date as revised: July 1993. Publication #1102.

Members of the American Institute of Chemical Engineers shall uphold and advance the integrity, honor and dignity of the engineering profession by: being honest and impartial and serving with fidelity their employers, their clients, and the public; striving to increase the competence and prestige of the engineering profession; and using their knowledge and skill for the enhancement of human welfare. To achieve these goals, members shall

American Institute of Chemical Engineers (AIChE)*

1. Hold paramount the safety, health and welfare of the public in performance of their professional duties.
2. Formally advise their employers or clients (and consider further disclosure, if warranted) if they perceive that a consequence of their duties will adversely affect the present or future health or safety of their colleagues or the public.
3. Accept responsibility for their actions and recognize the contributions of others; seek critical review of their work and offer objective criticism of the work of others.
4. Issue statements or present information only in an objective and truthful manner.
5. Act in professional matters for each employer or client as faithful agents or trustees, and avoid conflicts of interest.
6. Treat fairly all colleagues and co-workers, recognizing their unique contributions and capabilities.
7. Perform professional services only in areas of their competence.
8. Build their professional reputations on the merits of their services.
9. Continue their professional development throughout their careers, and provide opportunities for the professional development of those under their supervision.

---

* 1992 version. Reprinted with permission of AIChE.

American Society of Civil Engineers (ASCE)*

Fundamental Principles

Engineers uphold and advance the integrity, honor and dignity of the engineering profession by:

1. using their knowledge and skill for the enhancement of human welfare;
2. being honest and impartial and serving with fidelity the public, their employers and clients;
3. striving to increase the competence and prestige of the engineering profession; and
4. supporting the professional and technical societies of their disciplines.

Fundamental Canons

1. Engineers shall hold paramount the safety, health and welfare of the public in the performance of their professional duties.
2. Engineers shall perform services only in areas of their competence.
3. Engineers shall issue public statements only in an objective and truthful manner.
4. Engineers shall act in professional matters for each employer or client as faithful agents or trustees, and shall avoid conflicts of interest.
5. Engineers shall build their professional reputation on the merit of their services and shall not compete unfairly with others.
6. Engineers shall act in such a manner as to uphold and enhance the honor, integrity, and dignity of the engineering profession.
7. Engineers shall continue their professional development throughout their careers, and shall provide opportunities for the professional development of those engineers under their supervision.

ASCE Guidelines to Practice under the Fundamental Canons of Ethics

Canon 1.

Engineers shall hold paramount the safety, health and welfare of the public in the performance of their professional duties.

a. Engineers shall recognize that the lives, safety, health and welfare of the general public are dependent upon engineer-

---

* 1998 version. Reprinted with permission of ASCE.

ing judgments, decisions and practices incorporated into structures, machines, products, processes and devices.

b. Engineers shall approve or seal only those design documents, reviewed or prepared by them, which are determined to be safe for public health and welfare in conformity with accepted engineering standards.

c. Engineers whose professional judgment is overruled under circumstances where the safety, health and welfare of the public are endangered, shall inform their clients or employers of the possible consequences.

d. Engineers who have knowledge or reason to believe that another person or firm may be in violation of any of the provisions of Canon 1 shall present such information to the proper authority in writing and shall cooperate with the proper authority in furnishing such further information or assistance as may be required.

e. Engineers should seek opportunities to be of constructive service in civic affairs and work for the advancement of the safety, health and well-being of their communities.

f. Engineers should be committed to improving the environment to enhance the quality of life.

Canon 2.

Engineers shall perform services only in areas of their competence.

a. Engineers shall undertake to perform engineering assignments only when qualified by education or experience in the technical field of engineering involved.

b. Engineers may accept an assignment requiring education or experience outside of their own fields of competence, provided their services are restricted to those phases of the project in which they are qualified. All other phases of such project shall be performed by qualified associates, consultants, or employees.

c. Engineers shall not affix their signatures or seals to any engineering plan or document dealing with subject matter in which they lack competence by virtue of education or experience or to any such plan or document not reviewed or prepared under their supervisory control.

Canon 3.

Engineers shall issue public statements only in an objective and truthful manner.

a. Engineers should endeavor to extend the public knowledge of engineering, and shall not participate in the dissemination of untrue, unfair or exaggerated statements regarding engineering.

   b. Engineers shall be objective and truthful in professional reports, statements, or testimony. They shall include all relevant and pertinent information in such reports, statements, or testimony.

   c. Engineers, when serving as expert witnesses, shall express an engineering opinion only when it is founded upon adequate knowledge of the facts, upon a background of technical competence, and upon honest conviction.

   d. Engineers shall issue no statements, criticisms, or arguments on engineering matters which are inspired or paid for by interested parties, unless they indicate on whose behalf the statements are made.

   e. Engineers shall be dignified and modest in explaining their work and merit, and will avoid any act tending to promote their own interests at the expense of the integrity, honor and dignity of the profession.

Canon 4.

Engineers shall act in professional matters for each employer or client as faithful agents or trustees, and shall avoid conflicts of interest.

   a. Engineers shall avoid all known or potential conflicts of interest with their employers or clients and shall promptly inform their employers or clients of any business association, interests, or circumstances which could influence their judgment or the quality of their services.

   b. Engineers shall not accept compensation from more than one party for services on the same project, or for services pertaining to the same project, unless the circumstances are fully disclosed to and agreed to, by all interested parties.

   c. Engineers shall not solicit or accept gratuities, directly or indirectly, from contractors, their agents, or other parties dealing with their clients or employers in connections with work for which they are responsible.

   d. Engineers in public service as members, advisors, or employees of a governmental body or department shall not participate in considerations or actions with respect to services solicited or provided by them or their organization in private or public engineering practice.

   e. Engineers shall advise their employers or clients when, as a result of their studies, they believe a project will not be successful.

   f. Engineers shall not use confidential information coming to them in the course of their assignments as a means of making personal profit if such action is adverse to the interests of their clients, employers or the public.

g. Engineers shall not accept professional employment outside of their regular work or interest without the knowledge of their employers.

Canon 5.

Engineers shall build their professional reputation on the merit of their services and shall not compete unfairly with others.

a. Engineers shall not give, solicit or receive either directly or indirectly, any political contribution, gratuity, or unlawful consideration in order to secure work, exclusive of securing salaried positions through employment agencies.

b. Engineers should negotiate contracts for professional services fairly and on the basis of demonstrated competence and qualifications for the type of professional service required.

c. Engineers may request, propose or accept professional commissions on a contingent basis only under circumstances in which their professional judgments would not be compromised.

d. Engineers shall not falsify or permit misrepresentation of their academic or professional qualifications or experience.

e. Engineers shall give proper credit for engineering work to those to whom credit is due, and shall recognize the proprietary interests of others. Whenever possible, they shall name the person or persons who may be responsible for designs, inventions, writings or other accomplishments.

f. Engineers may advertise professional services in a way that does not contain misleading language or is in any other manner derogatory to the dignity of the profession. Examples of permissible advertising are as follows:

Professional cards in recognized, dignified publications, and listings in rosters or directories published by responsible organizations, provided that the cards or listings are consistent in size and content and are in a section of the publication regularly devoted to such professional cards.

Brochures which factually describe experience, facilities, personnel and capacity to render service, providing they are not misleading with respect to the engineer's participation in projects-described.

Display advertising in recognized dignified business and professional publications, providing it is factual and is not misleading with respect to the engineer's extent of participation in projects described.

A statement of the engineers' names or the name of the firm and statement of the type of service posted on projects for which they render services.

Preparation or authorization of descriptive articles for the lay or technical press, which are factual and dignified. Such articles shall not imply anything more than direct participation in the project described.

Permission by engineers for their names to be used in commercial advertisements, such as may be published by contractors, material suppliers, etc., only by means of modest, dignified notation acknowledging the engineers' participation in the project described. Such permission shall not include public endorsement of proprietary products.

g. Engineers shall not maliciously or falsely, directly or indirectly, injure the professional reputation, prospects, practice or employment of another engineer or indiscriminately criticize another's work.

h. Engineers shall not use equipment, supplies, laboratory or office facilities of their employers to carry on outside private practice without the consent of their employers.

Canon 6.

Engineers shall act in such a manner as to uphold and enhance the honor, integrity, and dignity of the engineering profession.

a. Engineers shall not knowingly act in a manner which will be derogatory to the honor, integrity, or dignity of the engineering profession or knowingly engage in business or professional practices of a fraudulent, dishonest or unethical nature.

Canon 7.

Engineers shall continue their professional development throughout their careers, and shall provide opportunities for the professional development of those engineers under their supervision.

a. Engineers should keep current in their specialty fields by engaging in professional practice, participating in continuing education courses, reading in the technical literature, and attending professional meetings and seminars.

b. Engineers should encourage their engineering employees to become registered at the earliest possible date.

c. Engineers should encourage engineering employees to attend and present papers at professional and technical society meetings.

d. Engineers shall uphold the principle of mutually satisfying relationships between employers and employees with respect to terms of employment including professional grade descriptions, salary ranges, and fringe benefits.

The Fundamental Principles

Engineers uphold and advance the integrity, honor, and dignity of the Engineering profession by:

I. using their knowledge and skill for the enhancement of human welfare;

II. being honest and impartial, and serving with fidelity the public, their employers and clients, and

III. striving to increase the competence and prestige of the engineering profession.

The Fundamental Canons

1. Engineers shall hold paramount the safety, health and welfare of the public in the performance of their professional duties.

2. Engineers shall perform services only in the areas of their competence.

3. Engineers shall continue their professional development throughout their careers and shall provide opportunities for the professional and ethical development of those engineers under their supervision.

4. Engineers shall act in professional matters for each employer or client as faithful agents or trustees, and shall avoid conflicts of interest or the appearance of conflicts of interest.

5. Engineers shall build their professional reputation on the merit of their services and shall not compete unfairly with others.

6. Engineers shall associate only with reputable persons or organizations.

7. Engineers shall issue public statements only in an objective and truthful manner.

8. Engineers shall consider environmental impact in the performance of their professional duties.

The ASME Criteria for Interpretation of the Canons

The ASME criteria for interpretation of the Canons are advisory in character and represent the objectives toward which members of the engineering profession should strive. They constitute a body of principles upon which an engineer can rely for guidance in specific situations. In addition, they provide interpretive guidance to the ASME Board on Professional Practice and Ethics in applying the code of Ethics of Engineers.

---

* 1999 version. Fundamental Canons and Principles as of 1999. Reprinted with permission of ASME.

1. Engineers shall hold paramount the safety, health and welfare of the public in the performance of their professional duties.

   a. Engineers shall recognize that the lives, safety, health and welfare of general public are dependent upon engineering judgments, decisions and practices incorporated into structures, machines, products, processes and devices.

   b. Engineers shall not approve or seal plans and/or specifications that are not of a design safe to the public health and welfare and in conformity with accepted engineering standards.

   c. Whenever the Engineers' professional judgment is overruled under circumstances where the safety, health, and welfare of the public are endangered, the Engineers shall inform their clients and/or employers of the possible consequences.

      (1) Engineers shall endeavor to provide data such as published standards, test codes, and quality control procedures that will enable the users to understand safe use during life expectancy associated with the designs, products, or systems for which they are responsible.

      (2) Engineers shall conduct reviews of the safety and reliability of the designs, products, or systems for which they are responsible before giving their approval to the plans for the design.

      (3) Whenever Engineers observe conditions, directly related to their employment, which they believe will endanger public safety or health, they shall inform the proper authority of the situation.

   d. If engineers have knowledge of or reason to believe that another person or firm may be in violation of any of the provisions of these Canons, they shall present such information to the proper authority in writing and shall cooperate with the proper authority in furnishing such further information or assistance as may be required.

2. Engineers shall perform services only in areas of their competence.

   a. Engineers shall undertake to perform engineering assignments only when qualified by education and/or experience in the specific technical field of engineering involved.

   b. Engineers may accept an assignment requiring education and/or experience outside of their own fields of competence, but their services shall be restricted to other phases of the project in which they are qualified. All other phases of such project shall be performed by qualified associates, consultants, or employees.

3. Engineers shall continue their professional development throughout their careers, and should provide opportunities for the professional and ethical development of those engineers under their supervision.

4. Engineers shall act in professional matters for each employer or client as faithful agents or trustees, and shall avoid conflicts of interest or the appearance of conflicts of interest.

   a. Engineers shall avoid all known conflicts of interest with their employers or clients and shall promptly inform their employers or clients of any business association, interests, or circumstances which could influence their judgment or the quality of their services.

   b. Engineers shall not undertake any assignments which would knowingly create a potential conflict of interest between themselves and their clients of their employers.

   c. Engineers shall not accept compensation, financial or otherwise, from more than one party for services on the same project, or for services pertaining to the same project, unless the circumstances are fully disclosed to, and agreed to, by all interested parties.

   d. Engineers shall not solicit or accept financial or other valuable considerations, for specifying products or material or equipment suppliers, without disclosure to their clients or employers.

   e. Engineers shall not solicit or accept gratuities, directly or indirectly, from contractors, their agents, or other parties dealing with their clients or employers in connection with work for which they are responsible.

   f. When in public service as members, advisors, or employees of a governmental body or department, Engineers shall not participate in considerations or actions with respect to services provided by them or their organization(s) in private or product engineering practice.

   g. Engineers shall not solicit an engineering contract from a governmental body on which a principal, officer, or employee of their organization serves as a member.

   h. When, as a result of their studies, Engineers believe a project(s) will not be successful, they shall so advise their employer or client.

   i. Engineers shall treat information coming to them in the course of their assignments as confidential, and shall not use such information as a means of making personal profit if such action is adverse to the interests of their clients, their employers or the public.

i.1 They will not disclose confidential information concerning the business affairs or technical processes of any present or former employer or client or bidder under evaluation, without his consent, unless required by law.

i.2 They shall not reveal confidential information or finding of any commission or board of which they are members unless required by law.

i.3 Designs supplied to Engineers by clients shall not be duplicated by the Engineers for others without the express permission of the client(s).

j. The Engineer shall act with fairness and justice to all parties when administering a construction (or other) contract.

k. Before undertaking work for others in which the Engineer may make improvements, plans, designs, inventions, or other records which may justify seeking copyrights or patents, the Engineer shall enter into a positive agreement regarding the rights of respective parties.

l. Engineers shall admit their own errors when proven wrong and refrain from distorting or altering the facts to justify their decisions.

m. Engineers shall not accept professional employment outside of their regular work or interest without the knowledge of their employers.

n. Engineers shall not attempt to attract an employee from another employer by false or misleading representations.

5. Engineers shall build their professional reputation on the merit of their services and shall not compete unfairly with others.

a. Engineers shall negotiate contracts for professional services on the basis of demonstrated competence and qualifications for the type of professional service required and at fair and reasonable prices.

b. Engineers shall not request, propose, or accept professional commissions on a contingent basis under circumstances under which their professional judgments may be comprised.

c. Engineers shall not falsify or permit misrepresentation of their, or their associates, academic or professional qualification. They shall not misrepresent or exaggerate their degrees of responsibility in or for the subject matter of prior assignments. Brochures or other presentations incident to the solicitation of employment shall not misrepresent pertinent facts concerning employers, employees, associates, joint ventures, or their past accomplishments.

    d.   Engineers shall prepare only articles for the lay or technical press which are factual, dignified and free from ostentation or laudatory implications. Such articles shall not imply other than their direct participation in the work described unless credit is given to others for their share of the work.

    e.   Engineers shall not maliciously or falsely, directly or indirectly, injure the professional reputation, prospects, practice or employment of another engineer, nor shall they indiscriminately criticize another's work.

    f.   Engineers shall not use equipment, supplies, laboratory or office facilities of their employers to carry on outside private practice without consent.

6. Engineers shall associate only with reputable persons or organizations.

    a.   Engineers shall not knowingly associate with or permit the use of their names or firm names in business ventures by any person or firm which they know, or have reason to believe, are engaging in business or professional practices of a fraudulent or dishonest nature.

    b.   Engineers shall not use association with non-engineers, corporations, or partnerships to disguise unethical acts.

7. Engineers shall issue public statements only in an objective and truthful manner.

    a.   Engineers shall endeavor to extend public knowledge, and to prevent misunderstandings of the achievements of engineering.

    b.   Engineers shall be completely objective and truthful in all professional reports, statements or testimony. They shall include all relevant and pertinent information in such reports, statements, or testimony.

    c.   Engineers, when serving as expert or technical witnesses before any court, commission, or other tribunal, shall express an engineering opinion only when it is founded upon adequate knowledge of the facts in issue, upon a background of technical competence in the subject matter, and upon honest conviction of the accuracy and propriety of their testimony.

    d.   Engineers shall issue no statements, criticisms, or arguments on engineering matters which are inspired or paid for by an interested party, or parties, unless they preface their comments by identifying themselves, by disclosing the identities of the party or parties on whose behalf they are speaking, and by revealing the existence of any pecuniary interest they may have in matters under discussion.

    e.   Engineers shall be dignified and modest in explaining their work and merit, and shall avoid any act tending to promote their own interest at the expense of the integrity, honor and dignity of the profession or another individual.

8. Any Engineer accepting membership in The American Society of Mechanical Engineers by this action agrees to abide by this Society Policy on Ethics and procedures for implementation.

Institute of Electronic and Electrical Engineers (IEEE)*

We, the members of the IEEE, in recognition of the importance of our technologies in affecting the quality of life throughout the world, and in accepting a personal obligation to our profession, its members and the communities we serve, do hereby commit ourselves to the highest ethical and professional conduct and agree:

1. to accept responsibility in making engineering decisions consistent with the safety, health, and welfare of the public, and to disclose promptly factors that might endanger the public or the environment;
2. to avoid real or perceived conflicts of interest whenever possible, and to disclose them to affected parties when they do exist;
3. to be honest and realistic in stating claims or estimates based on available data;
4. to reject bribery in all its forms;
5. to improve the understanding of technology, its appropriate application, and potential consequences;
6. to maintain and improve our technical competence and to undertake technological tasks for others only if qualified by training or experience, or after full disclosure of pertinent limitations;
7. to seek, accept, and offer honest criticism of technical work, to acknowledge and correct errors, and to credit properly the contributions of others;
8. to treat fairly all persons regardless of such factors as race, religion, gender, disability, age, or national origin;
9. to avoid injuring others, their property, reputation, or employment by false or malicious action;
10. to assist colleagues and co-workers in their professional development and to support them in following this code of ethics.

Effective January 1, 1991

Adopted by the Board of Directors, March 8, 1912.

A. General Principles.

B. The Engineer's Relations to Client or Employer.

C. Ownership of Engineering Records and Data.

D. The Engineer's Relations to the Public.

E. The Engineer's Relations to the Engineering Fraternity.

F. Amendments.

American Institute of Electrical Engineers (AIEE)*

While the following principles express, generally, the engineer's relations to client, employer, the public, and the engineering fraternity, it is not presumed that they define all of the engineer's duties and obligations.

A. General Principles

1. In all of his relations the engineer should be guided by the highest principles of honor.

2. It is the duty of the engineer to satisfy himself to the best of his ability that the enterprises with which he becomes identified are of legitimate character. If after becoming associated with an enterprise he finds it to be of questionable character, he should sever his connection with it as soon as practicable.

B. The Engineer's Relations to Client or Employer

3. The engineer should consider the protection of a client's or employer's interests his first professional obligation, and therefore should avoid every act contrary to this duty. If any other considerations, such as professional obligations or restrictions, interfere with his meeting the legitimate expectation of a client or employer, the engineer should inform him of the situation.

4. An engineer cannot honorably accept compensation, financial or otherwise, from more than one interested party, without the consent of all parties. The engineer, whether consulting, designing, installing or operating, must not accept commissions, directly or indirectly, from parties dealing with his client or employer.

5. An engineer called upon to decide on the use of inventions, apparatus, or anything in which he has a financial interest, should make his status in the matter clearly understood before engagement.

---

* Trans. AIEE, v. 31, part 2, pp. 2227–2230, 1912. Also found in *Turning Points in American Electrical History* by J. E. Brittain (New York: IEEE Press, 1976).

6. An engineer in independent practice may be employed by more than one party, when the interests of the several parties do not conflict; and it should be understood that he is not expected to devote his entire time to the work of one, but is free to carry out other engagements. A consulting engineer permanently retained by a party, should notify others of this affiliation before entering into relations with them, if, in his opinion, the interests might conflict.

7. An engineer should consider it his duty to make every effort to remedy dangerous defects in apparatus or structures or dangerous conditions of operation, and should bring these to the attention of his client or employer.

C. Ownership of Engineering Records and Data

8. It is desirable that an engineer undertaking for others work in connection with which he may make improvements, inventions, plans, designs, or other records, should enter into an agreement regarding their ownership.

9. If an engineer uses information which is not common knowledge or public property, but which he obtains from a client or employer, the results in the form of plans, designs, or other records, should not be regarded as his property, but the property of his client or employer.

10. If an engineer uses only his own knowledge, or information which by prior publication, or otherwise, is public property and obtains no engineering data from a client or employer, except performance specifications or routine information; then in the absence of an agreement to the contrary the results in the form of inventions, plans, designs, or other records, should be regarded as the property of the engineer, and the client or employer should be entitled to the use only in the case for which the engineer was retained.

11. All work and results accomplished by the engineer in the form of inventions, plans, designs, or other records, that are outside of the field of engineering for which a client or employer has retained him, should be regarded as the engineer's property unless there is an agreement to the contrary.

12. When an engineer or manufacturer builds apparatus from designs supplied to him by a customer, the designs remain the property of the customer and should not be duplicated by the engineer or manufacturer for others without express permission. When the engineer or manufacturer and a customer jointly work out designs and plans or develop inventions a clear understanding should be reached before the beginning of the work regarding the respective rights of

ownership in any inventions, designs, or matters of similar character, that may result.

13. Any engineering data or information which an engineer obtains from his client or employer, which he creates as a result of such information, must be considered confidential by the engineer; and while he is justified in using such data or information in his own practice as forming part of his professional experience, its publication without express permission is improper.

14. Designs, data, records and notes made by an employee and referring exclusively to his employer's work, should be regarded as his employer's property.

15. A customer, in buying apparatus, does not acquire any right in its design but only the use of the apparatus purchased. A client does not acquire any right to the plans made by a consulting engineer except for the specific case for which they were made.

D. The Engineer's Relations to the Public

16. The engineer should endeavor to assist the public to a fair and correct general understanding of engineering matters, to extend the general knowledge of engineering, and to discourage the appearance of untrue, unfair or exaggerated statements on engineering subjects in the press or elsewhere, especially if these statements may lead to, or are made for the purpose of, inducing the public to participate in unworthy enterprises.

17. Technical discussions and criticisms of engineering subjects should not be conducted in the public press, but before engineering societies, or in the technical press.

18. It is desirable that first publication concerning inventions or other engineering advances should not be made through the public press, but before engineering societies or through technical publications.

19. It is unprofessional to give an opinion on a subject without being fully informed as to all the facts relating thereto and as to the purposes for which the information is asked. The opinion should contain a full statement of the conditions under which it applies.

E. The Engineer's Relations to the Engineering Fraternity

20. The engineer should take an interest in and assist his fellow engineers by exchange of general information and expe-

rience, by instruction and similar aid, through the engineering societies or by other means. He should endeavor to protect all reputable engineers from misrepresentation.

21. The engineer should take care that credit for engineering work is attributed to those who, so far as his knowledge of the matter goes, are the real authors of such work.

22. An engineer in responsible charge of work should not permit nontechnical persons to overrule his engineering judgments on purely engineering grounds.

F. Amendments

Additions to, or modifications in, this Code may be made by the Board of Directors under the procedure applying to a by-law.

Index

Key: (#):   a number within parentheses appearing right after a page number indicates an item (e.g. discussion topic) that is being referred to on that page

    (f#):   indicates citation is a footnote or literary source

    italics: indicates a case study, discussion topic, or vehicle name (such as orbiter *Challenger*)